Gateway Routing Selection Schemes for Post-Disaster Recovery in Mobile Ad Hoc Networks

Gateway Routing Selection Schemes for Post-Disaster Recovery in Mobile Ad Hoc Networks bridges the gap by providing practical guidelines for an efficient design and performance evaluation of gateway selection schemes to manage load balancing. This book provides both a theoretical background and a practical evaluation of gateway selection methods using simulation methodology.

- Provides good coverage in a single text on a performance evaluation of gateway routing selection schemes for post-disaster recovery

- Offers students, teachers, and researchers both theoretical and practical knowledge of system design and performance validation

- Enhances teaching and learning and research capability in gateway routing selection schemes

- Begins each chapter with a set of learning objectives and provides chapter summaries as well as review questions

- Provides illustrations, mini-projects, and a list of acronyms

Gateway Routing Selection Schemes for Post-Disaster Recovery in Mobile Ad Hoc Networks makes the teaching, learning, and researching of gateway routing selection schemes a more active process by using practical tools and exercises.

Gateway Routing Selection Schemes for Post-Disaster Recovery in Mobile Ad Hoc Networks

Nor Aida Mahiddin and Nurul I. Sarkar

CRC Press
Taylor & Francis Group
Boca Raton London New York

CRC Press is an imprint of the
Taylor & Francis Group, an **informa** business

A CHAPMAN & HALL BOOK

First edition published 2025
by CRC Press
2385 Executive Center Drive, Suite 320, Boca Raton, FL 33431

and by CRC Press
4 Park Square, Milton Park, Abingdon, Oxon, OX14 4RN

CRC Press is an imprint of Taylor & Francis Group, LLC

© 2025 Nor Aida Mahiddin and Nurul I. Sarkar

ISBN: 978-1-032-70054-0 (hbk)
ISBN: 978-1-032-70055-7 (pbk)
ISBN: 978-1-032-70057-1 (ebk)

DOI: 10.1201/9781032700571

Typeset in Minion
by codeMantra

Contents

Preface

THERE HAVE been many natural and human-made disasters that have occurred, destroying urban and rural areas in recent years worldwide. It can be devastating when existing tele-communication infrastructure is destroyed by disasters. A Mobile Ad hoc Network (MANET) provides a solution to the problem of connectivity in disaster affected areas. This MANET has the potential to provide quick network connectivity that does not require any infrastructure. However, in disaster recovery areas, MANET performance can be deteriorated due to network traffic congestion. The high traffic generated by hand-held and mobile devices of victims and their loved ones can create a bottleneck at the MANET gateways.

A good knowledge of MANET gateway routing selection schemes for post-disaster recovery are required for the efficient design and deployment of such systems. While there are numerous literature and textbooks on MANET gateway routings, very few of them present systematic studies addressing the issues impacting MANET performance. This book bridges the gap by providing practical guidelines for an efficient design and performance evaluation of gateway selection scheme to manage the load balancing. This book provides both theoretical background and practical evaluation of gateway selection scheme using simulation methodology.

This book has the following main features:

- Provides good coverage in a single text on evaluation of gateway routing selection for post-disaster recovery for students and network researchers from industry and academia.

- Offers students, teachers, researchers, and network designers both theoretical and practical knowledge of design and performance evaluation of gateway routing selection schemes.

- Enhances teaching and learning, and research capability in gateway routing selection schemes at all levels.

- Begins each chapter with a set of learning objectives and provides chapter summary as well as review questions at the end.

- Provides illustrations, mini-projects and a list of acronyms.

ORGANIZATION AND OUTLINE

This book is organized into six chapters.

Chapters 1 and 2 provide theoretical background for gateway routing selection schemes for post-disaster recovery in Mobile Ad hoc networks. Chapter 1 provides a rationale for this book. Chapter 2 reports on the design of the proposed gateway selection scheme, and it provides a detailed evaluation of this scheme's performance using the OMNET++ simulation tool. The simulation results are presented as empirical evidence, confirming the superior performance of the proposed scheme in terms of critical metrics such as packet throughput, packet delay, packet drop ratio, packet delivery ratio, and sent packet rate.

Chapters 3–5 collectively constitute the core contributions of this book. These chapters are primarily concerned with quantifying the factors that impact MANET performance in disaster recovery environments. Simulation methodologies are employed to rigorously evaluate network performance, with the simulated disaster area in Loja City, Ecuador, serving as a representative environment for mimicking real disaster scenarios. In Chapter 3, the focus shifts to investigating the impact of routing selection schemes on MANET performance within disaster recovery contexts. The primary objective is to compare the performance of the proposed routing scheme against traditional routing schemes like AODV and DSDV. Chapter 4 provides system performance study by simulation. Results are presented through various studies, including developing a simulation model to study the performance of a typical gateway routing selection scheme, investigating the effect of node mobility on the system performance, and evaluating the performance of the routing schemes for AODV and DSDV routing protocols.

Chapter 5 provides advanced concepts and future directions in Mobile Ad hoc Networks MANETs) for disaster recovery. It explores future trends, especially integrating emerging network technologies such as Wi-Fi 6, 5G, and Internet of Things (IoT). The energy-efficient network, and advanced routing techniques including machine learning for dynamic route optimization are also explored. Finally, Chapter 6 summarizes the key findings from Chapters 4 and 5 and provides implications for deployment of MANETs in disaster recovery.

TARGET AUDIENCE FOR THIS BOOK

This book is written for college and university students, teachers, and professional audiences. This book would be a useful resource for undergraduate, post-graduate and research students, teachers, and network engineers who are interested in getting insight into MANET design and performance evaluation in post-disaster recovery scenarios. It should also be useful to practitioners working in the wireless networking and telecommunication industry.

LEARNING AIDS

This book provides the following learning aids:

- **Learning Objectives:** Each chapter begins with a list of learning objectives that preview the chapter's main ideas and highlights the key concepts and skills that students can achieve by completing the chapter.

- **Figures:** The key concept in MANET gateway routing selections is illustrated using diagrams throughout this book. These illustrations help students to develop a better understanding of MANET gateway routing concepts.

- **Summary:** Each chapter provides a summary of the contents presented in the chapter. This helps students to preview key ideas in the chapter before moving on to the subsequent chapters.

- **Key Terms:** Each chapter provides a set of key terms and abbreviations. Both students and teachers can benefit by using the listing of key terms to recall MANET gateway concepts before and after reading the chapter.

- **Review Questions**: Each chapter provides a set of review questions linked to the learning objectives, allowing the teachers to evaluate their teaching effectiveness. Teachers can use the review questions to initiate class discussion.

- **Mini Project:** For each chapter, a list of projects provided for students to explore key MANET routing concepts and to gain a deeper understanding of the topics and solutions described in the chapter. The practical activities in the form of mini projects can be carried out by simulation.

Acknowledgments

WE THANK the authority of Universiti Sultan Zainal Abidin (Malaysia) and Auckland University of Technology (New Zealand) for their support throughout the project. Our thanks go to the entire production team at Taylor & Francis Group for their ongoing support. Lastly and most importantly we would like to thank our family members for their patience, love, and encouragement throughout this project.

Nor Aida Mahiddin and Nurul I. Sarkar

Authors

Nor Aida Mahiddin earned a BS in information technology at the National University of Malaysia, a master's in computer science with a major in distributed computing, and a PhD in computer and information sciences at Auckland University of Technology, New Zealand. She is a Senior Lecturer in the Faculty of Informatics and Computing at Universiti Sultan Zainal Abidin, Malaysia, and is affiliated with the East Coast Environmental Research Institute (ESERI).

Dr. Mahiddin is the author of numerous papers published in peer-reviewed journals and conference proceedings. She is also an active member of professional organizations, including the Institute of Electrical and Electronics Engineers (IEEE), the Internet Society, and the Society of Digital Information and Wireless Communications (SDIWC).

Her research interests span a diverse range of areas with a focus on disaster recovery planning and infrastructure. Additionally, she explores emergency communication frameworks and their optimization as well as ad hoc network design and traffic flow frameworks.

 **Nurul I. Sarkar** is a Professor and the Director of the Networking and Security Research Centre at Auckland University of Technology (AUT), New Zealand. He is a keynote speaker, chair, and committee member for various national and international fora. Professor Sarkar has 30 years of teaching experience in universities at both undergraduate and postgraduate levels and has taught a range of subjects, including next-generation networking, computing technologies in society, Cisco networking, IoT, data communications, and wireless networks. He earned a PhD in electrical, computer, and software engineering (wireless networks) at the University of Auckland. *Improving the Performance of Wireless LANs: A Practical Guide* (Taylor & Francis, 2014) is his second book. His first book, *Tools for Teaching Computer Networking and Hardware Concepts*, was published by IGI Global in 2006 and has received commendation worldwide.

Professor Sarkar has published over 200 articles (16+ Q1 journals since 2018) in refereed international journal and conference proceedings, including *IEEE Communications Magazine, IEEE Internet of Things Journal, IEEE Transactions on Vehicular Technology, IEEE Transactions on Network and Service Management, IEEE Transactions on Education, Ad Hoc Networks, Computer Communications, Journal of Network and Computer Applications, Computer Networks,* and *Sensors.* He has spent periods of research leave in China, Japan, and Malaysia in recent years. He was a conference general Co-Chair for ITNAC'19 and CECNet'18 and TPC Co-Chair for ICOIN'24 and IEEE ICC14. He was the co-recipient of the 2017 Best Paper Award from the 31st ICOIN for a paper on mobility-aware network selection for V2I Communication over LTE-A networks. Professor Sarkar is a Member of ACM and a Senior Member of IEEE Communications and Vehicular Technology Societies and the Australasian Association for Engineering Education. His research interests include wireless network protocols, cognitive radio ad hoc networks, IoT, and fog computing.

Acronyms

A4LP	A4LP Routing Protocol
ABR	Associativity-Based Routing
ANSI	Ad Hoc Networking with Swarm Intelligence
AODV	Ad Hoc On-Demand Distance Vector
AOMDV	Ad Hoc On-Demand Multipath Distance Vector Routing
AQOR	Ad Hoc QoS On-Demand Routing
ARA	The Ant-Colony Based Routing Algorithms
Beraldi	Polarized Gossip Protocol for Path Discovery
CGSR	Cluster Head Gateway Switch Routing
DAR	Distributed Ant Routing
DBR2P	Dynamic Backup Routes Routing Protocol
DDR	Distributed Dynamic Routing
DSDV	Destination-Sequenced Distance Vector
DSR	Dynamic Source Routing
DST	Distributed Spanning Tree
FORP	The Flow Oriented Routing Protocol
FSR	Fisheye State Routing
FZRP	Fisheye Zone Routing Protocol
GRP	Gathering Based Routing Protocol
GSR	Global State Routing
GWRS	Gateway and Routing Selection
HOLSR	A Hierarchical Proactive Routing Mechanism for Mobile Ad Hoc Networks
HOPNET	Hybrid Ant Colony Optimization

IG	Internet Gateway
LANMAR	Landmark Ad Hoc Routing
LDR	Labelled Distance Routing
LMST	Local Minimum Spanning Tree
LRHR	Link Reliability-Based Hybrid Routing
LSR	Labelled Successor Routing
LTRT	Local Tree-Based Reliable Topology
MANET	Mobile Ad hoc Network
MN	Mobile Node
OD-PFS	On-Demand Packet Forwarding Scheme
OLSR	Optimized Link State Routing
QMRB	QoS Routing with Traffic Distribution
QOLSR	OLSR with Quality of Service
QoS	Quality of Service
RDMAR	Relative Distance Micro-Discovery Ad Hoc Routing
R-DSDV	Randomized Destination-Sequenced Distance Vector
ROAM	Routing On-Demand Acyclic Multipath
RPR	Recycled Path Routing
RREQ	Route Request
SLR	Source Routing with Local Recovery
SLURP	Scalable Location Update-Based Routing Protocol
SMORT	Scalable Multipath On-Demand Routing
SSBR	Signal Stability-Based Adaptive Routing
SSL	Source-Sequenced labels
STAR	Source-Tree Adaptive Routing
SWORP	Stable Weight Based On-Demand Routing Protocol
TCP	Transmission Control Protocol
TORA	Temporally Ordered Routing Algorithm
WRP	Wireless Routing Protocol
ZHLS	Zone-Based Hierarchical Link State Routing Protocol
ZRP	Zone Routing Protocol

CHAPTER 1

Introduction

LEARNING OBJECTIVES

After reading and completing this chapter, you will be able to:

- Discuss the architecture of a typical MANET.
- List and describe three potential applications of MANETs.
- Discuss how MANET can be used in disaster recovery.
- Discuss why load balancing is important in MANETs.

1.1 MOBILE AD HOC NETWORKS

The expansion of wireless technology has revolutionized data transmission through radio waves. In this network paradigm, nodes communicate with each other without reliance on fixed station access points. The network's structure can spontaneously form and reconfigure without the need for a central network system. Mobile Ad Hoc Networks (MANETs) have gained widespread popularity with the advent of smart computing devices and the advancement of wireless communications.

Routing is the process of transmitting information from a source to a destination within an internetwork. To achieve

DOI: 10.1201/9781032700571-1

this, data packets are forwarded from the source to the destination through neighboring nodes, often referred to as routers. These routers play a crucial role in forwarding data packets across multiple hops until they reach their intended destination. Notably, the topology of a MANET is highly dynamic and unpredictable [1].

In the context of wireless communications, the IEEE 802.11 standard governs operations and defines two primary modes: infrastructure-based and infrastructure-less (ad hoc). Infrastructure-based modes function as Wi-Fi hotspots, enabling devices to connect to the Internet through access points. However, in dynamic environments where devices or users connect temporarily, the infrastructure-less or ad hoc mode proves more efficient and flexible. In this mode, nodes form an independent basic service set or ad hoc network. After a synchronization phase, each node can communicate with the others.

Moreover, if one of the nodes, referred to as node A, in the ad hoc network is connected to a wired network, it assumes the role of a gateway, granting wireless Internet access to all nodes within the ad hoc network. This gateway node facilitates seamless communication between the ad hoc network and the broader Internet. In practical communication scenarios, ad hoc networks often serve as mobile users' communication networks, allowing users' devices to function as network infrastructure components, including routers, switches, and servers, as long as they remain within transmission range.

To further illustrate the structure of a MANET, Figure 1.1 provides a visual representation of its layered architecture. It comprises various components, including applications for sending messages, individual devices serving as MANET nodes, and some nodes within the network's coverage area acting as gateways to manage traffic. To facilitate communication among all MANET nodes, protocols like Wi-Fi Direct are necessary.

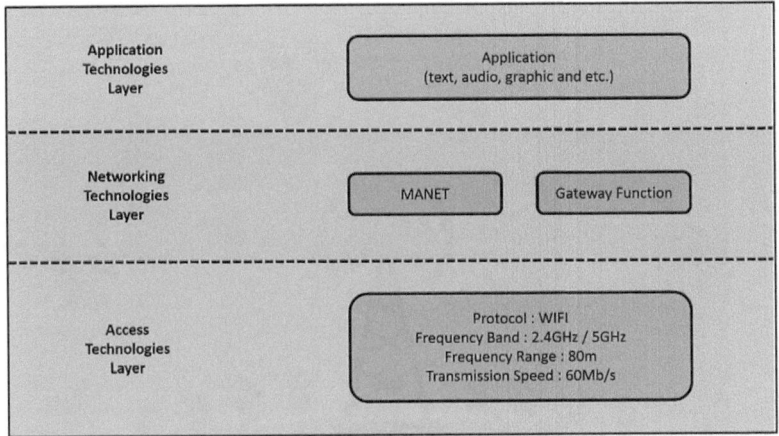

FIGURE 1.1   A typical MANET architecture.

1.2 APPLICATIONS OF MANET

MANET were initially proposed for military applications and have since found utility in various fields. Among these applications, deploying MANETs in disaster scenarios presents a significant challenge, but it can be an invaluable solution [2]. After a disaster strikes, the rapid collection and exchange of information about victims are crucial for effective rescue operations. For instance, earthquakes, tsunamis, floods, and other natural disasters like those in Great East Japan, Wenchuan, China, Indonesia, and Malaysia have had devastating impacts on communities. Figure 1.2 illustrates the breakdown in communication and information exchange that often occurs during such disasters.

Collapsed structures resulting from earthquakes can leave disaster victims trapped without the means to make voice phone calls using mobile devices [3]. As depicted in Figure 1.2, mobile devices cease to function when the communication infrastructure in the area is compromised. MANETs, being decentralized and infrastructure-independent, offer a viable solution to these communication challenges [4]. Their unique capabilities enable

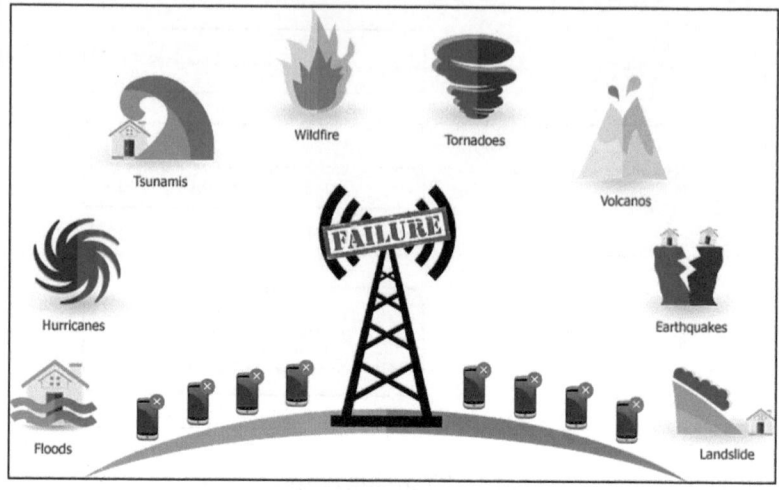

FIGURE 1.2   Failure in communication and information exchange.

users to establish dynamically reconfigurable wireless networks without relying on fixed infrastructure, making them highly valuable [5].

Table 1.1 lists a MANET application and their corresponding scenarios [6, 7]. The evolution of mobile computing equipment and network communication infrastructure has transformed the way people communicate, retrieve information, and carry out tasks. Individuals now use mobile devices to stay connected and exchange information, regardless of location or the presence of fixed network infrastructure. MANETs facilitate information exchange anytime and anywhere, contributing to the flexibility and resilience of modern communication systems.

1.2.1 MANET in Disaster Recovery Areas

In a MANET network, each node is limited to communicating only with other nodes within the same network. When disaster victims seek to establish communication with the outside world to inform their family and friends of their safety, their objective is to send messages, share vital information, or even transmit

TABLE 1.1   Applications of MANETs

Application Area	Scenario and Potential Services
Military communication and operations	• Keep the communication networks of soldiers, vehicles, and military always in a good condition and ensure they stay connected
Disaster scenario	• Emergency rescue operation takes over the communication when existing communication infrastructure has been damaged or cut off for a safety reason. Generally, it can be used in rescue operations to support medic teams such as earthquake, flood, disaster relief, etc.
Commercial sectors	• Shopping malls • Airports • Sport stadiums • E-commerce • Vehicular Ad Hoc network
Home networking	• Indoor and outdoor internet access • Personal area networks
Enterprise networking	• Indoor and outdoor internet access • Conferences • Meeting rooms
Education	• Virtual classrooms • Ad Hoc communication through meetings or lectures
Sensor networks	• Smart home applications: smart sensors for home appliances • Geo-location tracking device for humans or animals
Entertainment	• Multi-user games • Robotic pets

videos. However, it's essential to note that these messages must find a way to reach the broader Internet.

During a disaster event, it's a common occurrence for critical infrastructure to become compromised. Power outages, server failures, and communication service disruptions can render devices like cell phones, iPads, or laptops useless for disaster victims attempting to communicate with their loved ones. However, we are fortunate to live in a technological era that has provided solutions for accessing energy in post-disaster situations. For instance, during the 2015 earthquake in Nepal with a magnitude of 7.8, the non-profit organization SunFarmer supplied solar power systems and batteries to remote hospitals and schools, as well as repaired streetlights in the city [8, 9]. Solar power played a crucial role in Nepal's disaster relief efforts.

It's worth noting that while this research recognizes the critical importance of energy in disaster scenarios, addressing energy-related challenges after a disaster is not within the scope of this study. Instead, this study is focused on addressing the issue of establishing network connections for mobile devices to access the Internet once power-related challenges have been resolved. Furthermore, in situations where Internet connectivity is absent, all devices equipped with wireless networking capabilities can spontaneously form a network to exchange information. This capability is made possible by the decentralized nature of such networks, which do not rely on traditional infrastructure. Consequently, MANETs emerge as an ideal solution for addressing these communication challenges during disaster events.

Hence, as depicted in Figure 1.3, this book introduces the Gateway Routing Selection (GWRS) Scheme to improve network performance for Internet access in disaster recovery areas. As

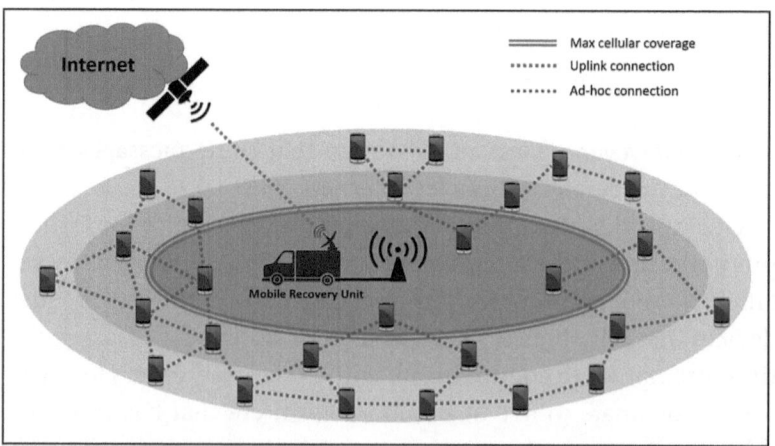

FIGURE 1.3   Internet connectivity through MANETs.

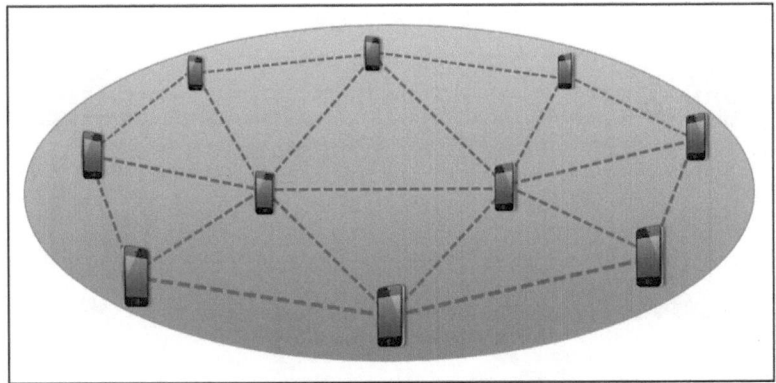

FIGURE 1.4   MANET connectivity in disaster areas.

previously mentioned, this research operates under the assumption of a post-disaster scenario where electricity sources have been either maintained or have a backup power supply. Under these conditions, mobile devices with Internet coverage are designated as the MANET gateway. Devices lacking Internet coverage will route their packets through neighboring nodes until the packets reach the designated MANET gateway.

Under normal circumstances, many devices rely on established communication infrastructure. However, in the aftermath of a disaster, network devices become isolated due to the collapse and damage of communication infrastructure. To expand communication capabilities, as illustrated in Figure 1.4, the use of MANET technologies allows nodes to continue communicating. For instance, if the destination node falls outside the transmission range of the source node, a neighboring node will function as a relay, forwarding the message until it reaches the intended destination (as seen in Figure 1.5). It's important to note that within a MANET, each node can only communicate with others within the same network. In the scenario presented, we assume the presence of three MANET nodes within Internet range. These three nodes serve as gateways, enabling other nodes in the network to connect to the Internet.

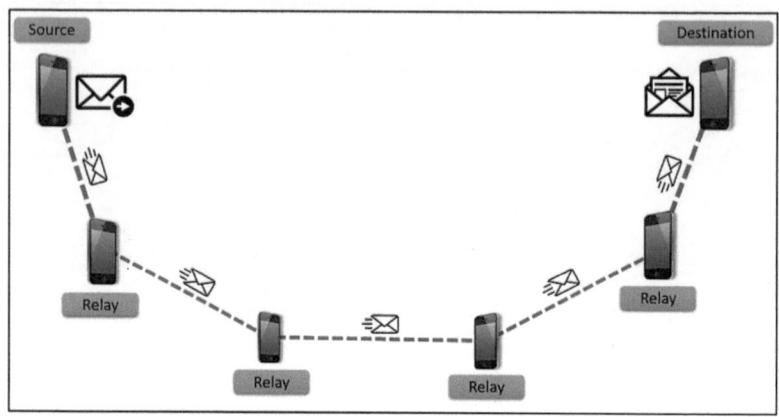

FIGURE 1.5   Neighbor node as a relay.

1.2.2 Gateway Load Balancing

Numerous studies have explored the concept of gateways within MANETs. Gateways serve as entry points for nodes within a MANET to establish connections with external networks. In disaster scenarios where communication networks fail and take time to be restored, MANETs provide an immediate solution. However, the primary challenge in such situations is the potential heavy traffic load. Consequently, the concept of gateway load balancing becomes crucial in preventing congestion. When there are multiple gateways within a network, load balancing among them must be carefully considered to enhance network performance. Miao et al. [10], for instance, introduced an intelligent selection gateway scheme aimed at identifying overloaded gateway nodes.

Prior research [11] has proposed several techniques to mitigate congestion through load balancing, including the Queue-Based Load Balancing Algorithm, Aggregate Interface Queue Length, and the Hybrid Registration Mechanism. The objective of a gateway load balancing scheme is to distribute tasks evenly among all gateways, thereby achieving load equalization. These techniques are typically enhancements of the Ad Hoc On-Demand Distance Vector (AODV) routing protocol method.

The process of gateway selection plays a pivotal role in connecting communication nodes between infrastructure networks and non-infrastructure networks, such as the Internet and MANETs. Traditionally, the shortest path selection is based on minimizing hop count, which serves as a relay for traffic between the MANET and the Internet. Several studies have focused on improving individual network performance aspects, such as network throughput, end-to-end delay, or packet delivery ratio. Bouk et al. [12] proposed a comprehensive gateway selection scheme that enhanced overall network performance, accuracy in path availability computation, path load capacity, and latency. Manoharan and Mohanalakshmie [13] introduced a trust-based hybrid gateway selection scheme, emphasizing the importance of choosing paths through trusted nodes and uncongested routes to reach the network gateway. They incorporated security elements like node trust, route trust, and residual route load capacity. Zaman et al. [11] conducted similar integrated research but placed specific focus on gateway load balancing strategies, drawing inspiration from AODV to develop a new gateway selection strategy designed to evenly distribute packets among network gateways.

Many of the proposed techniques observed are modifications of conventional routing protocol methods like AODV and DSDV. Tashtoush et al. [6], for instance, introduced a method that employed hop count as a weight value for gateways. It limited the number of routes using Fibonacci weights to choose the most efficient route based on hop count. However, the continuous computation of route weight introduced network overhead.

To enhance communication in disaster-stricken areas, Liu et al. [7] presented a method to identify gateway nominees without imposing a heavy computational load. However, this method considered only a single channel and a solitary gateway for a given area. The selection of different gateways can impact network performance, particularly in terms of throughput.

Prabhavat et al. [14] evaluated various load distribution methods across a multipath network using different criteria. They emphasized the significance of splitting traffic and route selection as the primary techniques for achieving improved load balancing performance in the network. However, their study did not delve into routing methods for establishing multiple paths. Nevertheless, efficient techniques for maintaining packet order and preventing packet loss remain crucial, as they can maximize throughput performance.

1.3 CONTRIBUTION AND STRUCTURE OF THIS BOOK

The overall structure of this book is shown in Figure 1.6. This book's developmental phase is centered around the exploration of gateway and routing selection schemes.

Chapters 2 and 3 serve as a foundational and informative backdrop for the subsequent discussions in this book. To achieve accurate performance evaluation, a solid foundation in MANETs

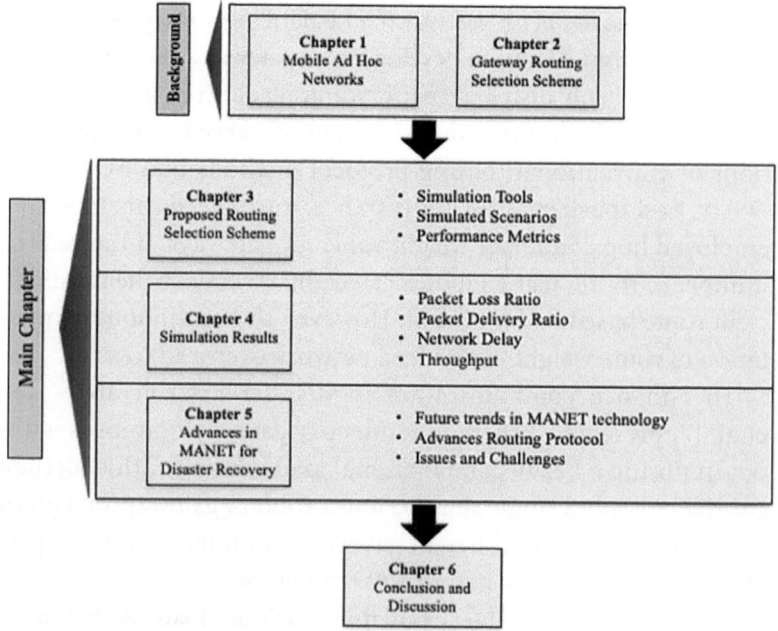

FIGURE 1.6 The structure of this book.

and the dynamics of disaster recovery scenarios is essential. This understanding encompasses aspects like post-disaster network architecture, traffic loads at gateways, route discovery processes during disaster recovery, and the utilization of performance evaluation tools. Building a robust understanding of gateway and routing selection schemes within MANETs and disaster recovery scenarios is a crucial prerequisite for enhancing network performance.

Chapter 2 delves into the realm of MANET's gateway and routing selection schemes in disaster recovery situations. It places significant emphasis on elucidating the architecture of MANETs, the diverse applications of MANETs, and the prevailing issues and challenges faced in disaster recovery areas. Furthermore, this chapter identifies and describes simulation tools used to evaluate performance in this context. The selection of appropriate MANET performance metrics is also addressed, with a focus on metrics such as packet throughput, end-to-end packet delay, packet sending rates, and packet loss ratios, which are instrumental in validating the effectiveness of proposed schemes. The performance of MANETs in disaster recovery scenarios is profoundly influenced by gateway and routing schemes, a subject thoroughly examined in Chapter 3. This chapter is dedicated to an in-depth exploration of the intricacies surrounding the design of gateway and routing selection schemes. It undertakes a comprehensive survey of previous research efforts related to design and performance enhancement within this domain.

Chapter 4 offers a comprehensive presentation of the major findings derived from the proposed selection scheme, with insights into network planning, deployment strategies, and the evolution of MANETs in disaster recovery areas.

In Chapter 5, we provide advanced concepts and future research directions in MANETs focusing on disaster recovery scenarios. The future directions focus on integration of emerging technologies including 5G, Wi-Fi 6, and Internet of Things, energy-efficiency, advanced routing techniques, and machine learning for dynamic route optimization. The book culminates in

Chapter 6, where a concise summary and concluding remarks are presented. Additionally, it outlines potential future developments within this research domain.

1.4 SUMMARY

In this chapter, we have established the fundamental principles and necessary background information concerning MANETs. We have also shed light on the specific scenario and conditions prevalent in disaster recovery areas. Moreover, we have delved into the challenges associated with mitigating network communication failures during such catastrophic events. To explore potential solutions without the need for costly real-size disaster scenario experiments, we have conducted a thorough examination of MANET simulation tools as a practical and viable alternative.

1.5 KEY TERMS

MANET, Applications, Infrastructure-Based Network, Infrastructure-Less Network (Ad Hoc), IEEE 802.11 Standard, Wi-Fi Direct, Disaster Recovery, Network Topology, Gateway Node, Internet Access, Communication Infrastructure, Energy Sources, Network Performance, Load Balancing, Traffic Load, Routing Protocol, Hop Count, Multipath Network, Simulation Tools.

1.6 REVIEW QUESTIONS

1. Describe the architecture of a typical MANET in disaster area. Your answer must include a diagram to illustrate the situation.

2. MANET can be useful to set up a quick communication link and services in disaster areas. List and describe three practical applications of MANETs.

3. Discuss how MANET can be used in disaster recovery situations.

4. Discuss the importance of load balancing in MANETs.

5. Network researchers are facing real challenges in improving the performance of a typical MANET. List and describe two potential performance issues/challenges in MANET design.

1.7 MINI PROJECTS

The following mini projects aim to provide a deeper understanding of the topics covered in this chapter through a review of literature. More mini projects have been presented in the subsequent chapters of this book to emphasize hands-on practical learning experience.

1. Network researchers are facing real challenges in improving the performance of MANETs in recent years. To gain an insight into the performance of MANET, it is useful to be able to review a literature on various issues and challenges related to MANET performance. Conduct an in-depth literature review on MANET performance issues and challenges.

2. People can be benefited from MANETs specially in disaster areas. This mini project aims to enhance the knowledge and understanding of MANET architecture, protocols, and applications. Conduct an in-depth review of literature on MANETs. You can use Table 1.2 as a template to record your findings.

TABLE 1.2   Leading Researchers and their Contributions to MANET Architecture and Protocols

Researcher	Contribution	Year	Description/Key Concept

CHAPTER 2

Gateway Routing Selection Schemes

LEARNING OBJECTIVES

After reading and completing this chapter, you will be able to:

- Identify and discuss two main challenges in the design of MANET gateway routing selection schemes.

- Identify and discuss two main features of MANET routing selection schemes.

- Discuss the complexity of route selection in congested networks.

- Discuss the effect of node mobility in disaster recovery scenarios.

2.1 INTRODUCTION

This chapter discusses the crucial role of gateways in facilitating communication between Mobile Ad Hoc Network (MANET) nodes and the Internet. These gateways serve as essential entry

   DOI: 10.1201/9781032700571-2

and exit points for network packets. Within a MANET, specific nodes are designated as Internet gateways (IGs), responsible for routing packets to and from the Internet. Their primary function is to manage the flow of network traffic between multiple distinct networks, and it is worth noting that a single network can incorporate multiple gateways. Each gateway is equipped with a mechanism to monitor the average queue size.

In disaster recovery scenarios, individuals seek to establish connections to the Internet, resulting in a surge of data traffic directed towards the same gateway. In this context, the efficient preservation of packet order becomes paramount to prevent packet loss. Effective routing techniques play a pivotal role in maximizing throughput performance. Routing aims to determine the optimal path for packet transmission, typically measured based on various routing parameters, such as distance, bandwidth, delay, and load [1]. However, it's essential to acknowledge that routing within mobile wireless networks poses considerable challenges.

2.2 PREVIOUS RESEARCH ON MANET GATEWAYS

Recent research has placed significant focus on gateways, which serve as the entry points allowing nodes within a MANET to establish connections with external networks. There are scenarios, especially in the aftermath of disasters, where communication networks fail, and it takes time to restore them. In such situations, MANETs can offer an immediate solution. However, a major challenge arises due to the substantial surge in traffic as individuals seek to contact their family and friends.

To restore communication effectively in disaster-stricken areas, where many victims simultaneously attempt to connect to the Internet, an organized gateway selection scheme becomes imperative. This scheme helps manage packet congestion at each gateway and optimizes network throughput. The primary objective of gateway load balancing schemes is to distribute tasks evenly among all gateways, achieving load equalization. Various techniques have been proposed to alleviate congestion through load balancing, including the Queue-Based Load Balancing Algorithm, Aggregate Interface Queue Length, and the Hybrid Registration Mechanism. These techniques build upon the Ad Hoc On-Demand Distance Vector (AODV) routing protocol method, enhancing gateway load balancing's critical role in averting congestion.

In pursuit of continuous Quality of Services (QoSs) and Quality of Experience delivery, the MONET approach employs a hybrid MANET-Satellite network, offering a solution for the challenges posed by highly mobile, dynamic, and remote environments [15].

2.3 ISSUES IN GATEWAY SELECTION SCHEMES

A gateway, which is a mobile device with connectivity to external networks, plays a crucial role in influencing MANET performance. The potential for congestion arises when the volume of packets being sent to the gateway surpasses a predetermined threshold queue size. This can lead to significant packet loss, resulting in a substantial degradation of packet throughput performance. In situations where only one gateway is available within a MANET, all packets are directed through this single gateway to reach the Internet. In such instances, the queue size can quickly become overwhelmed when all network nodes simultaneously transmit packets through the same gateway. Consequently, the implementation of multiple gateways emerges as the solution to tackle this challenge. However, effectively distributing the traffic load among these gateways becomes a critical issue [16].

2.3.1 Broadcast

Based on the gateway literature discussed, prior schemes have employed a method involving the transmission of gateway broadcasting advertisement messages to all nodes within the network, essentially soliciting information regarding gateway coordinates. These gateways would periodically dispatch advertisement messages, and intermediate nodes would subsequently rebroadcast these messages to their neighboring nodes. However, this approach resulted in packet flooding throughout the network and consequently escalated the traffic burden placed on each gateway.

2.3.2 Uneven Traffic Load

Gateway traffic load can fluctuate, occasionally exhibiting disparities due to varying user demands. In one scenario, there may be minimal user demands, while in another, a substantial influx of user demands may occur simultaneously. Even when employing multiple MANET gateways, the potential for uneven traffic distribution among these gateways remains if traffic management isn't executed efficiently.

2.3.3 Gateway Failure

In the scenario where only one gateway serves as the connection between MANET nodes and the external network, the entire network relies entirely on this singular gateway. Should this lone gateway encounter a failure, it would result in a complete disconnection of the entire network from the external network. However, in the presence of multiple gateways within the network, even if one gateway experiences a malfunction, nodes still retain the capability to access the external network. Nodes located within the range of non-functional gateways must reroute their connections to other available MANET gateways.

Our proposed scheme will primarily address three key aspects: first, the methodology for gateway selection based on the network's prevailing conditions; second, the implementation of stable load balancing measures among gateways; and third, the strategies for maintaining a high level of throughput throughout the network.

2.4 PREVIOUS RESEARCH ON MANET ROUTING

This section will delve into previous MANET routing schemes. Over the years, starting from the late 1990s, the literature has been teeming with novel routing protocols aimed at addressing the challenges posed by routing selection schemes. These challenges encompassed issues, such as packet broadcasting to the entire network, the existence of multiple routes from source to destination nodes, node mobility, and network overhead.

In 1997, routing protocols like Associativity-Based Routing (ABR) [17] and TORA [18] were developed with a primary objective of enhancing route stability and recovering from link failures in MANETs. ABR focused on enhancing route stability to reduce frequent route restarts, offering a loop-free protocol. Around the same time, WRP [19], based on a pathfinding algorithm, was introduced to mitigate looping in routing. This protocol is operated by distributing frequent routing update packets to maintain routing tables. Additionally, Chiang et al. [20] brought forward a cluster-based approach named Least Cluster Change (LCC). In LCC, nodes were organized into clusters, each under the governance of a cluster head. This clustering approach was aimed at resolving issues prevalent in large-scale MANETs. There were various clustering schemes tailored to different objectives, such as power efficiency, cost considerations, and connectivity concerns. However, a common challenge across these schemes was ensuring fair treatment for the cluster heads. This clustering approach, OD-PFS [21], adopted a blend of hierarchical and virtual backbone routing, where the network topology was mapped onto a virtual grid structure, followed by the division of nodes into clusters, each supervised by its own cluster head.

Moving into the 2000s, continuous routing methods began to emerge. LANMAR [22] and FSR [23] leveraged a fisheye technique to minimize routing overhead. In these methods, each node-maintained awareness of its neighbors within the surrounding fisheye scope area. Yang et al. [24] merged the routing approach of FSR with the zone routing protocol, where packets were transmitted between zone borders. ARA schemes [25] were also geared towards reducing routing overhead and took inspiration from ant foraging behaviors. These schemes employed the Ant technique for route discovery, which involved the propagation of forward and backward messages through flooding. The DAR [26] routing technique followed a similar Ant-inspired approach to mitigate computation complexity. The next hop was determined based on weight values, and ants would forward packets to choose the next hop with the highest probability. ANSI [27] introduced the utilization of Swarm Intelligence for routing selection, adapting this technique from the Ant routing method and maintaining multiple routes to the destination. Route discovery using these techniques aimed to find multiple routes, often as backups in case of route failure or disruption. Several of these routing schemes demonstrated that routing overhead could be reduced when nodes had alternative paths available to deliver packets to their destination. Even until 2009, the Ant technique persisted in various research endeavors. Wang et al. [28] emphasized the routing method HOPNET, which was founded on ZRP, Dynamic Source Routing (DSR), and the amalgamation of Ant Colony Optimization. This technique drew an analogy to ants hopping across zones, with forward ants gathering destination information from local nodes' routing tables. Then, the ants traversed between zones via border nodes. In the late 2000s, Reddy and Raghavan [29] devised a multipath routing protocol with one primary path and demonstrated that it could effectively reduce network overhead. The primary path was configured to be the shortest route, and nodes were permitted to receive multiple copies of the route request packet, although they were not allowed to reply to the source node, ultimately minimizing network overhead.

In the realm of network topology exploration, Chen et al. [30] maintained a comprehensive knowledge of the network topology, which included information about neighbors, next-hop nodes, and distances. The next-hop table contained an inventory of next-hop neighbors. A slew of studies in 1999 [31–34] proposed enhancements to broadcast methods. The AODV routing protocol aimed to reduce packet flooding across the network by introducing route information on demand, as opposed to continuously maintaining updated route information. However, LBAQ, introduced in 2007, still implemented flooding techniques using unique source-sequenced labels (SSLs) to disseminate Route REQuest messages (RREQs) within the network. Destination nodes were obligated to respond once they received these messages. The Relay Sequenced Label would only respond to unique RREP messages accompanied by SSLs to prevent path loops before the source node forwarded the packet along the route.

The ROAM and STAR schemes embraced an on-demand routing algorithm coupled with an enhanced topology broadcast protocol. Routes were established and sustained through diffusion computations to prevent routers from sending irrelevant packet requests to unreachable destinations. Each node possessed knowledge of a preferred link leading to potential destinations. FORP [35] schemes relied on a multi-hop handoff mechanism, where mobility prediction information was leveraged to anticipate topological changes. RDMAR [36] put forth an enhancement for route discovery and route maintenance through a loop-free routing protocol to minimize reactions when network topological changes occurred.

In contrast, AQOR [37] employed limited flooding during route discovery, with route requests incorporating bandwidth and end-to-end delay constraints. Messages were rebroadcast to the next hop if they satisfied these constraints. SLR [38] introduced a bypass routing technique aimed at improving the route discovery process in cases of link failures. This method triggered a local recovery process to bypass the broken link. Yu et al. explored the notion of replacing broken routes intelligently. They introduced a technique for dynamically switching out damaged routes, with intermediate nodes that overheard transmissions between source and destination nodes serving as potential candidates to replace failed nodes. DDR [39] is a contemporary solution that serves as a backup due to its high cost. It establishes connections only when needed and automatically terminates them when no data needs to be sent.

Many researchers have also made use of AODV routing protocols, illustrated by GRP [40]. In this scheme, source nodes broadcast destination query packets until they reach the destination node. Some source routing schemes, however, do not rely on a routing table [41]. For instance, DBR2P [42] operates without a routing table; source nodes acquire comprehensive route information directly from destination nodes. Multiple backup routes, determined by destination nodes with the assistance of intermediate nodes, are available in case of link failures. Even after nearly a decade, the issue of routing failure connections continued to be a focal point in numerous studies. SCaTR [43] put forth a solution: if a route to a destination was unavailable, a proxy request would be forwarded. When a proxy request message approached closer to the destination node, each node advertised itself as a proxy destination. Advertisement implied offering a route to the destination, while solicitation entailed requesting information from the destination [48]. In networks with asymmetric links, A4LP [44, 45], introduced a technique of limited packet forwarding. Receivers assessed a predefined fitness value with the sender before rebroadcasting a packet.

To date, several studies have incorporated weight mechanisms to manage route expiration times, error counts, and hop counts. Much like previous protocols, the source node initiated an RREQ message, and destination nodes responded with RREP messages. If multiple RREQ messages were received via different paths, the weight mechanism would evaluate the largest value to designate as the primary route [46]. LBAQ [47] measured link quality based on node mobility in the network, with link weight determined by availability, quality, and energy consumption. LRHR [48] established multiple routes from source to destination, with each route assigned an edge weight value, higher values indicating superior link reliability. QMRB [49] proposed a route selection scheme based on QoS considerations for each source-destination pair. The weight values for QoS encompassed factors like Static Resource Capacity, Dynamic Resource Availability, Neighborhood Quality, and Link Quality and Stability.

Numerous techniques have been built upon previous routing protocols. R-DSDV [50] introduced a randomized version of the DSDV protocol. This scheme employed a routing probability distribution technique where nodes adjusted their parameters to favor routes with lighter loads. OLSR [51] made use of hello and multipoint relays to discover route information, with source nodes not requiring the complete route information but possessing knowledge about the next hop. Building on this concept, HOLSR [52] introduced a hierarchical design, with multiple ad hoc networks (clusters) embedded within the larger network. Each cluster retained information about routing and nodes within its domain. QOLSR [53] proposed a multipoint relay selection process, augmenting OLSR routing techniques by considering delay and bandwidth parameters using three heuristics—QOLSR1, QOLSR2, and QOLSR3. In a bit difference method, John et al. [54] introduced the concept of distinguishing between "hot" and "cold" nodes in the forwarding strategy, where nodes closer to the destination were deemed "hotter" and responsible for rebroadcasting RREQ messages, while nodes with a "cold" value would discard the message. Beraldi et al. [55] leveraged meta-information to assist in packet forwarding by sending hints to neighboring nodes. Packets would be discarded by these neighbors if they did not exist.

Table 2.1 categorizes routing schemes based on several criteria, such as broadcast techniques, route metrics used in single or multiple routes between source and destination, the presence of a route depository in the form of a routing table or route cache, and the degree of communication complexity. This table provides an overview of prior research endeavors aimed at enhancing existing routing techniques.

Researchers have developed various routing schemes to explore optimal solutions in ad hoc routing. These ad hoc routing schemes can be categorized into three groups, as illustrated in Figure 2.1. The first category falls under the proactive approach, which relies on a routing table. A routing table comprises a set of rules organized in a table format, used to determine the path a data packet should take within the network. This approach is often referred to as "table-driven" routing. In table-driven routing, nodes are

TABLE 2.1 Optimisation of MANET Routing Schemes

Routing Scheme	Shortest Path	Broad-cast	Multiple Routes	Route Repository in RP	Over-head	Disaster Scenario	Node Mobility
ABR [17]	Strongest associativity	✓	-	✓	Medium	✗	✓
TORA [18]	✓	✓	✓	✓	High	✗	✗
SSBR [56]	Signal strength	✓	-	✓	Medium	✗	✓
FORP [57]	-	✓	-	✓	Medium	✗	✓
AODV [58]	✓	✓	-	✓	High	✗	✗
ROAM [32]	✓	-	✓	✓	Low	✗	✗
DSR [41]	✓	✓	✓	-	High	✗	✓
ARA [25]	✓	✓	✓	✓	Medium	✗	✓
AQOR [37]	Link bandwidth	✓	-	✓	Medium	✗	✓
DBR2P [42]	✓	-	-	-	-	✗	✓
RPR [54]	✓	✓	-	✓	High	✗	✗
GRP [40]	✓	✓	✓	-	High	✗	✓
SLR [38]	✓	-	✓	-	High	✗	✓
Beraldi [55]	-	✓	✓	-	High	✗	✓
LDR [59]	✓	-	-	✓	High	✗	✓
SMORT [29]	-	✓	-	-	-	✗	✓
Yu [60]	✓	-	-	-	-	✗	✓
LBAQ [47]	-	✓	✓	-	High	✗	✓

(Continued)

TABLE 2.1 Continued

Routing Scheme	Shortest Path	Broad-cast	Multiple Routes	Route Repository in RP	Over-head	Disaster Scenario	Node Mobility
LSR [61]	–	–	✓	✓	High	✗	✓
SWORP [46]	–	–	–	✓	High	✗	✓
OD-PFS [21]	–	–	–	✓	Medium	✗	✓
DAR [26]	Weighted	–	✓	✓	Medium	✗	✗
QMRB [49]	–	–	✓	✓	High	✗	✓
SCaTR [43]	–	–	✓	✓	High	✗	✓
DSDV [62]	✓	✓	–	✓	Low	✗	✗
WRP [19]	✓	–	–	✓	Low	✗	✗
CGSR [20]	✓	–	–	✓	Low	✗	✓
GSR [30]	✓	✓	–	✓	Low	✗	✓
STAR [33]	✓	–	–	✓	Low	✗	✓
R-DSDV [63]	✓	–	–	✓	Low	✗	✗
OLSR [51]	✓	–	–	✓	High	✗	✗
HOLSR [52]	✓	–	–	✓	High	✗	✓
QOLSR [53]	Periodic	–	–	✓	High	✗	✓
ZHLS [64]	✓	✓	✓	✓	Medium	✗	✓
DST [65]	Tree neighbor	–	✓	✓	Low	✗	✗
RDMAR [65]	✓	✓	–	✓	High	✗	✓
DDR [39]	Stable routing	–	✓	✓	Low	✗	✗
LANMAR [22]	✓	–	–	✓	Medium	✗	✓
FSR [23]	Scope range	–	–	✓	Low	✗	✓

Routing Scheme	Shortest Path	Broad-cast	Multiple Routes	Route Repository in RP	Over-head	Disaster Scenario	Node Mobility
SLURP [66]	InterZ / intraZ	-	✓	-	High	✗	✗
ZRP [67]	✓	-	-	-	-	✗	✓
ANSI [27]	✓	✓	✓	✓	Medium	✗	✓
FZRP [24]	✓	-	-	✓	Medium	✗	✓
A4LP [44] [45]	Power consumed	✓	✓	✓	Medium	✗	✓
HOPNET [28]	✓	-	-	✓	High	✗	✓
AOMDV [68]	-	✓	✓	✓	-	✗	✓
BATMAN [69]	✓	✓	-	-	Medium	✗	✗
BCHP [70]	-	✓	-	✓	-	✓	✓
DYMO [71]	✓	-	✓	✓	High	✗	✓

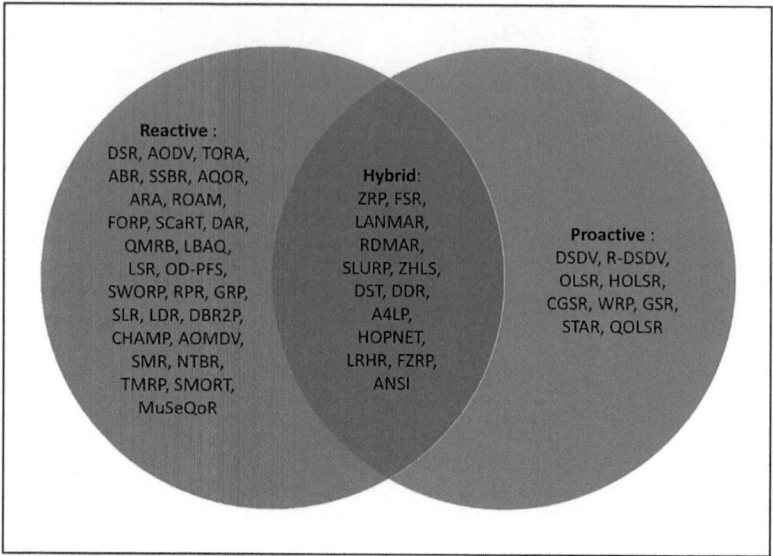

FIGURE 2.1   Routing schemes in MANETs.

constantly updated with routing information, regardless of when and how frequently such route details are needed. Routing information is stored within the routing table of each node within the network. However, maintaining an up-to-date routing table comes at the cost of increased communication overhead.

Proactive protocols, often referred to as table-driven protocols, continuously update routing information for all nodes within the network. This approach simplifies the task of source nodes in finding the path to the destination node since all paths are pre-established. This algorithm is stable and straightforward to implement in static network topologies. Local route calculations can be performed swiftly. However, in dynamic mobile environments, such as disaster recovery areas where each node represents a person and moves randomly, periodic updates for all nodes are inefficient. This approach leads to increased network overhead due to high channel usage. Additionally, nodes may join and leave the network to conserve battery life, making continuous routing information updates impractical. The benefit of this technique is that it can reduce waiting time for each node. Examples of well-known proactive routing protocols that have been used include OLSR (Optimized Link State Routing Protocol) and DSDV (Destination-Sequenced Distance Vector).

The second category encompasses reactive protocols. Reactive routing operates on-demand, providing route information from the sender to the receiver only when requested. Some reactive routing approaches have effectively reduced network traffic overhead. However, route discovery procedures, which often involve flooding the network through broadcast techniques, as seen in AODV, TORA, and DSR, are not efficient. AODV and DSR are examples of reactive algorithms. In reactive algorithms, source and destination routes are established only upon request, reducing network overhead. However, because the route is established only on-demand, the waiting time of nodes for route information is increased. This technique increases the total delay in the network. When a node wants to communicate with another node, two components are involved: (1) route discovery, to find a specific destination node location, and (2) route maintenance, to be able to manage route failure. One method for finding a destination node involves flooding the message throughout the entire network, which can work well with low traffic volumes but leads to network congestion and packet loss when traffic is high. In disaster recovery areas, the advantage of using a reactive protocol is that it enables energy saving during communication since there is no constant updating of table routing information, thus improving energy conservation.

AODV, which stands for Ad Hoc On-Demand Distance Vector Protocol, is a well-known reactive routing scheme employed in MANETs [58]. This scheme utilizes hop counting to determine the shortest path from the sender to the receiver. It is considered a reliable, straightforward, and effective metric. While it helps reduce routing overhead, it does pose a significant challenge in terms of packet delay because nodes need to wait for route connections to be established from the sender to the receiver. Routes are only established when specifically requested. This type of routing can be suitable for disaster scenarios where there are no obstacles in the path [72]. However, in situations where information is not continuously updated, there's a risk of communication loss if nodes suddenly vanish from the network. This is particularly problematic when the nodes within the network are highly mobile and rapidly changing their positions [73].

The hybrid routing protocol leverages the strengths of both proactive and reactive algorithms while aiming to mitigate the weaknesses inherent in each of them. This innovative approach achieves this by incorporating zone and cluster routing techniques. Hybrid routing is versatile and applicable across a broad spectrum of scenarios. Depending on specific network conditions, either reactive or table-driven approaches may be more suitable. Beyond the hybrid category, various other routing protocol categories exist, including location-aware, multipath, hierarchical, multicast, geographical multicast, and power-aware protocols. The hybrid category, which is the third category, combines the advantages of reactive and proactive protocols. Additionally, this hybrid routing scheme identifies specific zones within the network to minimize the excessive flooding of packets in a MANET when broadcasting messages [74]. Through this combination of techniques, overall network performance is enhanced.

Routing poses a significant challenge in mobile wireless networks, with its primary objective being the determination of the optimal path for packet transmission. To measure routing parameters, a standard algorithm takes into account factors like distance, bandwidth, delay, and load along a given path [1]. The concept of routing encompasses both a routing protocol or scheme and a routing algorithm. The role of a routing scheme is to facilitate the exchange of topological information and link weights, while the routing algorithm calculates the distance between nodes. The conventional algorithms used for calculating the shortest path include Dijkstra and Bellman-Ford's algorithms [75].

In the context of MANETs, routing schemes typically prioritize finding the shortest route from the sender to the receiver, assuming it's the optimal solution for achieving a high success rate within the network. However, these schemes often overlook the critical aspect of QoS assurance. QoS routing schemes, on the other hand, aim to identify one or more paths from a source to a destination, ensuring that the required bandwidth remains below the available total bandwidth. Adequate bandwidth is essential for seamless data transmission, particularly in real-time applications [76]. Examples of real-time applications encompass video conferencing, Voice over Internet Protocol, online gaming, and specific e-commerce transactions.

These applications demand functionalities that span a specified time frame, with response times typically needing to fall within certain maximum time limits, often measured in seconds. To cater to these requirements, a multipath QoS multicast routing protocol in a MANET was proposed in [76]. This protocol was designed to address the needs of QoS and bandwidth while transmitting data for real-time applications.

Among the well-known routing algorithms, Ant routing stands out. In [75], the performance of Ant routing in MANETs was analyzed, conducting simulations in two distinct scenarios: first, in a static network, and second, in dynamic network topologies. The performance results of AODV and DSR were used for comparison. The Ant routing algorithm demonstrated strong performance in static network topologies. However, its efficiency was compromised in dynamic topologies due to limited capacity and buffer size restrictions. In response to these challenges, ref [2] introduced a dynamic selection path method that depends on node and obstacle density. The author also considered issues related to destination selection, particularly in scenarios where multiple nodes might select the same destination based on node distances.

2.5 ISSUES IN ROUTING SCHEMES

A significant challenge associated with routing schemes in MANETs, as highlighted in [77], is the necessity for nodes to engage in route discovery processes to locate specific gateways for connecting to external networks. In disaster-prone areas, the nodes within the MANET can exhibit varying degrees of mobility, with some being stationary while others are mobile. Certain mobile nodes may be less dependable due to frequent device connections and disconnections driven by limited battery power. Additionally, routes can become unreliable after several transmission attempts.

In order to initiate route discovery within the network, nodes resort to broadcasting route requests, and destination nodes subsequently respond by transmitting route reply messages to nodes across the network. This approach leads to the flooding of the network with broadcast messages, a problem that significantly hampers the performance of MANETs. Consequently, it becomes imperative to incorporate congestion control mechanisms as a crucial network element to uphold the stability and reliability of the network.

2.5.1 Network Congestion

Network congestion arises when numerous mobile devices attempt to connect to the network simultaneously. To address this issue, research has established a predetermined allocation of data capacity for each gateway. However, without an effective routing selection scheme, congestion at each gateway cannot be alleviated.

2.5.2 Complexity of Route Selection

Traditional AODV routing protocols employ a broadcasting algorithm for route discovery when sending a packet to the gateway through multi-hop nodes. However, this approach involves broadcasting the packet to all nodes in the network, resulting in network flooding and routing complexity. The intricate route selection process causes delays in packet delivery to the destination. Therefore, this book will introduce an efficient routing selection scheme aimed at reducing delays in MANET performance. This scheme will streamline the route selection process to enhance network performance.

2.6 NODE MOBILITY IN DISASTER RECOVERY SCENARIOS

A MANET refers to a collection of mobile devices that can be established without the need for any pre-existing infrastructure. The simulation utilized a mobility model because it characterizes the behavior of these mobile nodes. In disaster scenarios, nodes within the network can exhibit high mobility due to panic reactions and efforts to escape from the disaster-affected area. Given this high mobility, there is a possibility that the destination node may suddenly move out of the coverage range. For instance, during the aftermath of the Great East Japan Earthquake, Suto et al. [78] observed that victims were moving around to seek reliable wireless access. Consequently, in a MANET, mobile devices often serve as both hosts and relays, necessitating adaptability to disruptions and changes in network topology because end-to-end connectivity relies on the positions and mobility of nodes [79]. Given the challenges posed by high mobility, the proposed scheme is designed to address scenarios in which the network environment undergoes dynamic changes [80].

2.7 TOPOLOGY RAPIDLY CHANGES

Disasters can result in power outages, necessitating efficient utilization of mobile devices by nodes in the network only when necessary. The network's topology undergoes alterations as nodes join and exit the network. Furthermore, some nodes are not stationary. They unpredictably change their positions at varying speeds within the disaster-stricken area. The combination of node mobility, node failures, nodes being powered off, and a range of node velocities presents significant challenges in maintaining communication within disaster recovery areas [77]. These challenges contribute to the creation of highly dynamic networks, which have a direct impact on the performance of MANETs.

2.8 SUMMARY

This chapter provided a comprehensive literature review, highlighting the key challenges related to gateways, routing, and node mobility in disaster recovery scenarios. These challenges are crucial to address to alleviate congestion at the gateway. The proposed scheme to tackle these issues was introduced, along with considerations regarding the routing selection scheme. Both aspects must be resolved to optimize MANET performance. Additionally, the chapter underscored the impact of node mobility in disaster areas, where network topology undergoes rapid changes, and the challenges posed by obstacles near nodes.

2.9 KEY TERMS

MANET, IG, Network Congestion, Routing Protocol, Node Mobility, Disaster Recovery Scenarios, Gateway Selection Scheme, Load Balancing, QoS, Hybrid MANET-Satellite Network, Broadcast, Proactive Routing Protocol, Reactive Routing Protocol, Hybrid Routing Protocol, Route Discovery.

2.10 REVIEW QUESTIONS

1. Explain the role of gateways in a MANET and their significance in disaster recovery scenarios.

2. What are the main challenges associated with gateway selection schemes in MANETs, especially in congested networks?

3. Describe the differences between proactive, reactive, and hybrid routing protocols in MANETs. Provide examples of each.

4. How does node mobility affect routing and communication in disaster recovery scenarios within a MANET?

5. Discuss the issues related to network congestion in MANETs and the importance of routing selection schemes in mitigating congestion.

2.11 MINI PROJECTS

The following mini projects are designed to provide hands-on experience in researching and understanding the concepts discussed in this chapter. This mini project is about "Investigation of Gateway Selection Schemes in MANETs."

Tasks: In this mini project, you will conduct an in-depth review of literature on gateway selection schemes in MANETs, with a particular focus on applications in disaster recovery scenarios. The objective is to develop a deeper understanding of methodologies and approaches used in designing and performance evaluation of gateway selection schemes. Follow the following steps to complete the mini project.

1. Literature review: Start by collecting research papers, articles, and books that discuss gateway selection schemes in MANETs. Focus on papers that address the challenges posed by disaster scenarios and network congestion. Make sure to include recent research.

2. Findings: Read and analyze the selected literature. Summarize the key findings related to gateway selection schemes, including the various approaches proposed to address congestion and improve network performance. Note any specific algorithms or techniques mentioned.

3. Comparison: Create a comparative analysis of different gateway selection schemes. Highlight the advantages and disadvantages of each approach, especially in the context of disaster recovery situations. Consider factors like scalability, resource efficiency, and reliability.

4. Open challenges: Based on your review, identify any open research challenges or areas where further improvements

are needed in gateway selection schemes for MANETs. Are there any emerging trends or technologies that could address these challenges?

5. Report: Compile your findings into a mini project report. Include a literature review section, a comparative analysis of gateway selection schemes, and a discussion of open challenges. Provide appropriate citations for your sources.

6. Presentation: Prepare a brief presentation summarizing your mini project findings. You can use slides to highlight key points and share your insights with others.

CHAPTER 3

Proposed Routing Selection Schemes

LEARNING OBJECTIVES

After reading and completing this chapter, you will be able to:

- Identify and explain the fundamental principles of AODV routing in MANETs.

- Explain the unique challenges associated with routing in emergency and disaster recovery scenarios, including network congestion, packet loss, and delays.

- Discuss, evaluate, and compare the performance of routing protocols, particularly AODV, DSDV, and a proposed routing selection scheme, in disaster recovery scenarios.

- Demonstrate proficiency in using simulation tools like OMNET++ for MANET performance assessment and understand their significance in practical scenarios.

- Calculate and interpret essential network performance metrics, such as end-to-end delay, packet loss ratio, and throughput, enabling you to assess and optimize MANETs effectively.

DOI: 10.1201/9781032700571-3

3.1 INTRODUCTION

AODV, which stands for Ad Hoc On-Demand Distance Vector, is a commonly employed routing scheme within Mobile Ad Hoc Networks (MANETs) [77]. This scheme relies on hop count as a metric to determine the shortest path between a sender and receiver. It is known for its simplicity and effectiveness, making it a trusted choice in MANETs. AODV employs a reactive routing approach, meaning that the route from sender to receiver is only established when needed. When a sender node wishes to establish a connection, it initiates a Route REQuest (RREQ) broadcast, and intermediate nodes forward this message until it reaches the destination node. However, this broadcast mechanism can lead to a broadcast storm, causing network inefficiency and flooding, as it sends messages to all nodes within range in search of the optimal route [72]. Each receiving node records temporary routes and selects the ones with the lowest hop count, reducing routing overhead. Nevertheless, a significant drawback of AODV is the delay incurred as nodes must wait for the route to be established from sender to receiver, which can be exacerbated when dealing with high-speed mobile devices within the network [73].

In our proposed scheme, which aims to address these issues, gateways within a MANET network proactively advertise their locations at regular intervals to all nodes within their range. Nodes receiving these advertisements store information about the nearest gateway in a routing table. When a node outside the gateway's range intends to send a message, other nodes in the network collaborate to forward the message until it reaches one of the gateways. However, if a gateway experiences heavy load conditions, a notification is broadcasted, prompting nearby nodes to seek an available alternative gateway. Therefore, this book employs a forward-and-backward technique to minimize packet loss. Recognizing that this technique may introduce packet delay, we have improved it by implementing an initial stage for heavy load notifications.

3.2 USE CASE SCENARIO

There exists a relatively limited body of literature that specifically addresses MANET routing protocols in emergency and rescue scenarios. An examination of routing schemes reveals that, among the reactive routing protocols, AODV demonstrates the best performance, while DSDV performs similarly well among proactive routing schemes within emergency scenarios. DSDV exhibits superior packet delivery performance. In our simulations, we compare the performance of our proposed routing selection scheme with that of DSDV and AODV, employing a case study set in a disaster recovery area. AODV and DSDV were selected for comparison as they excel in their respective routing protocol categories. Reina et al. [81] emphasized the significant impact of routing protocols on MANET performance in disaster scenarios, as they operate without the need for established infrastructures.

To conduct this comparative analysis, we utilized Loja City as a realistic environment for our case study, as previously discussed [72]. In scenario 2, we simulated our proposed scheme alongside AODV and DSDV routing schemes, configuring them under the same parameter environment. AODV relies on hop count to determine the shortest path from sender to receiver, and the route is only established when necessary. A sender node initiates a RREQ broadcast, which intermediate nodes forward until it reaches the destination node. However, this broadcasting technique can lead to a broadcast storm due to inefficiency, flooding the network by transmitting messages to all nodes within range in the quest for the optimal route. The act of broadcasting messages to discover a path to the destination contributes to increased network overhead. Each receiving node temporarily records potential routes, with preference given to routes with fewer hop counts.

In the context of disaster recovery areas, each node in the network represents an individual within the affected area, and these nodes are free to move randomly. Periodically updating routing information for each node in such a dynamic mobile environment proves inefficient and results in elevated network overhead due to high channel usage. Additionally, to conserve battery life, nodes in the network may frequently join and leave, making the constant refreshing of routing information ineffective in a highly mobile environment with a changing network topology.

The development of a routing scheme for MANETs in disaster recovery areas entails addressing several challenges, including (i) network congestion, (ii) node mobility, (iii) network overhead, and (iv) energy resources. It's worth noting that this book does not primarily focus on the energy problem but assumes that energy-related challenges have been resolved.

3.3 PROPOSED ROUTING SELECTION SCHEME IN DISASTER RECOVERY

This book introduces a routing selection scheme designed to streamline route selection processes effectively. The scheme intelligently manages the transmission of messages from nodes to gateways. To initiate the route from the sender to the receiver, nodes consult routing tables to determine available routes. In pursuit of an energy-efficient routing scheme, the proposed routing selection scheme updates the information in the routing table whenever new data becomes available, thereby enhancing energy conservation. The algorithm compares the current information with the previous data, and the routing table is updated if any discrepancies are detected. Figure 3.1 illustrates a flow chart outlining the proposed scheme. Before communication begins, as outlined in Algorithm 1, each gateway broadcasts its coordinates and current movement speed to neighboring nodes within a maximum transmission range. Each gateway is equipped with a predetermined threshold. When a gateway nears its capacity, it sends out full notifications to nodes at level one. The primary aim of this technique is to mitigate network congestion.

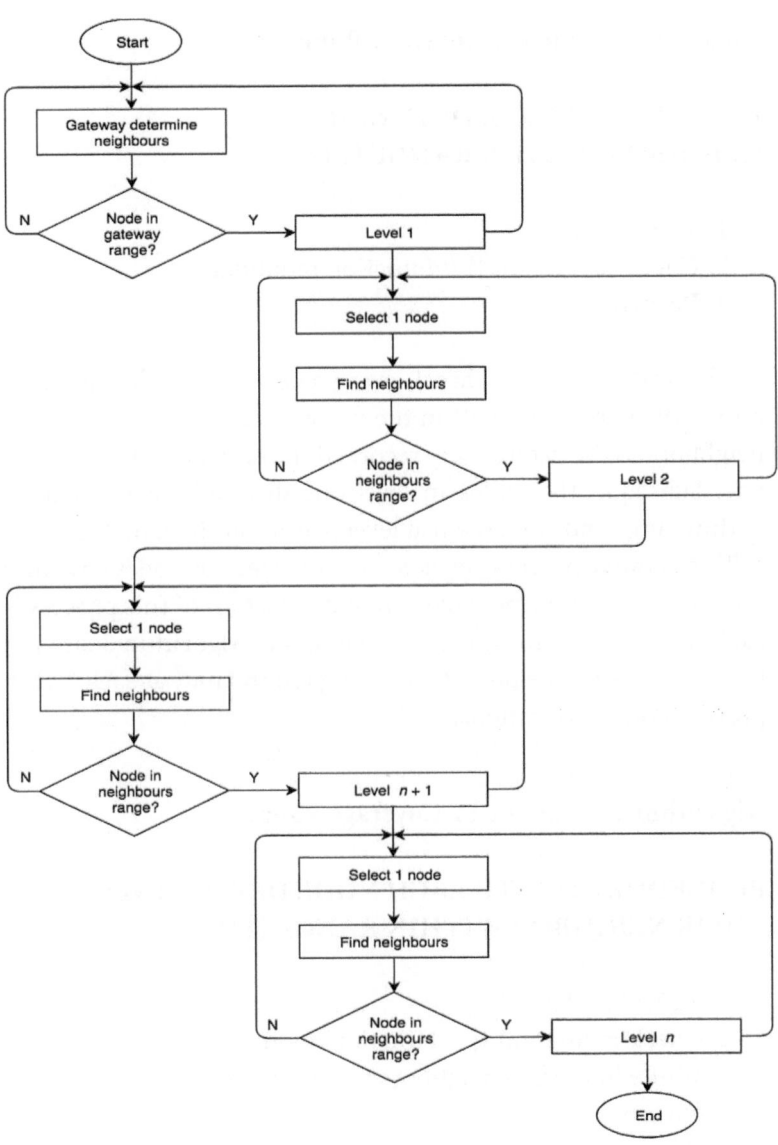

FIGURE 3.1 Flowchart of proposed efficient routing selection scheme.

Algorithm 1: Gateway Coverage Range

PROCEDURE FOR G: DETERMINES NEIGHBOR WITHIN RANGE (T1)

1: T1 ← Level 1
2: Check redundant {} // Function redundant
3: **Return**

As illustrated in Algorithm 1, every gateway identifies its neighbors, which are nodes within the gateway's coverage range. These neighbors of the gateway are recorded in the routing table at level one. Subsequently, nodes in level one discover their neighbors within range and store them at level two, as outlined in Algorithm 2. This iterative process persists until all neighbor nodes have been included in the routing table. Upon completion of this process by each node in determining their neighbors, Algorithm 3 simultaneously takes on the task of analyzing redundancy at each level to prevent node duplications.

Algorithm 2: Nodes in T1 Coverage Range

PROCEDURE FOR T1: EACH NODE DETERMINES THEIR NEIGHBOR WITHIN RANGE (T1)

1: T2 ← Level 2
2: Check redundant {} // Function Redundant
3: While find one S neighbors in upper level
4: Do send a packet
5: If U = 0
6: Then find one S neighbors in S level // No S neighbor at upper level
7: Upper level
8: N ← S neighbors

9: Send a packet
10: Else if waiting then
11: **Return**

Algorithm 3: Function to Check Redundancy

PROCEDURE TO COMPARE NODE LEVEL TN TO TN + 1

1: If at level Tn + 1 ← same nodes
   Then {
2: Remove the node
   }
3: **Return**

In accordance with Algorithm 4, when a node intends to transmit a packet beyond its local network, the source node initiates a RREQ directed to the gateway. The initial step involves inspecting the level of the source node. The process of identifying the next hop involves examining the source node's neighbors located at a higher level that also fall within the coverage range of the source node.

Algorithm 4: Source Node Generates an RREQ for the Internet via a Gateway

PROCEDURE TO CHECK THE LEVEL OF THE SOURCE NODE (S)

1: S ← sources node
2: U ← next hop // U = Upper level
3: While (U ≠ 0)
4: Do send a packet // Send to one neighbor only
5: **Return**

Algorithm 5: Route Discovery for RREQ

PROCEDURE TO CHECK THE LEVEL
OF THE SOURCE NODE (S)

1: S ← sources node
2: U ← next hop // U = Upper level
3: While find one S neighbors in upper level
4: Do send a packet
5: If U = 0 // No S neighbor at upper level
6: Then find one S neighbors in S level
7: N ← S neighbors
8: Send a packet
9: Else if waiting then
10: **Return**

In cases where there are no neighboring nodes in the upper level within the coverage range of the source node, the subsequent procedure, as outlined in Algorithm 5, comes into play. In this scenario, the RREQ is forwarded to another node within the network's coverage range, operating on the same level, with the aim of locating neighbors in the upper level. While this method may potentially introduce additional packet delay, it serves as a safeguard against packet loss, which is particularly critical in disaster recovery communication where the preservation of information is of utmost importance.

To provide a clearer understanding of the proposed schemes, Figure 3.1 illustrates the efficient routing selection scheme. The methodology entails gateways at the initial level of our routing scheme transmitting the packet out of the network, followed by the subsequent level comprising gateway neighbors. This sequential process continues until the last level of nodes is reached.

3.4 MODELING THE NETWORK

In the simulation, we employed the network area of Loja City as the simulation environment within OMNET++ (Objective Modular

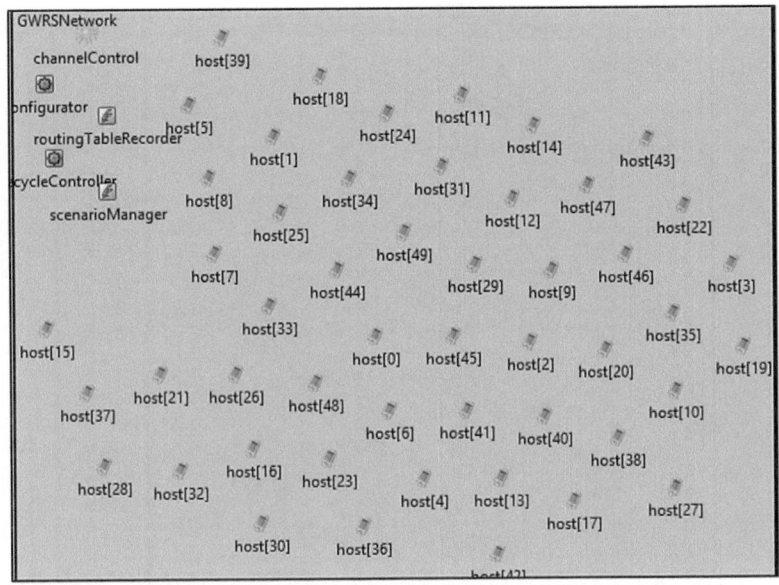

FIGURE 3.2   Omnet++ representation of minimum density in disaster area.

Network Testbed in C++). Figures 3.2 and 3.3 provide representations in OMNET++ depicting the minimum population density in the disaster area. For this simulation, we established a minimum of 50 mobile nodes (MNs). Conversely, Figures 3.4 and 3.5 illustrate the OMNET++ representations showcasing the maximum population density within the disaster area, with a maximum of 200 nodes defined.

The GWRS network, as depicted in the top-left box, governed the behavior of all nodes within the simulation model. Channel control was responsible for defining the channels used and determining the signal transmission range for each individual node. Meanwhile, the routing table recorder diligently recorded the flow of the routing scheme throughout the simulation.

3.4.1 Simulation Tools

In assessing the performance of a multi-hop wireless network, various methods can be employed, including analytical modeling,

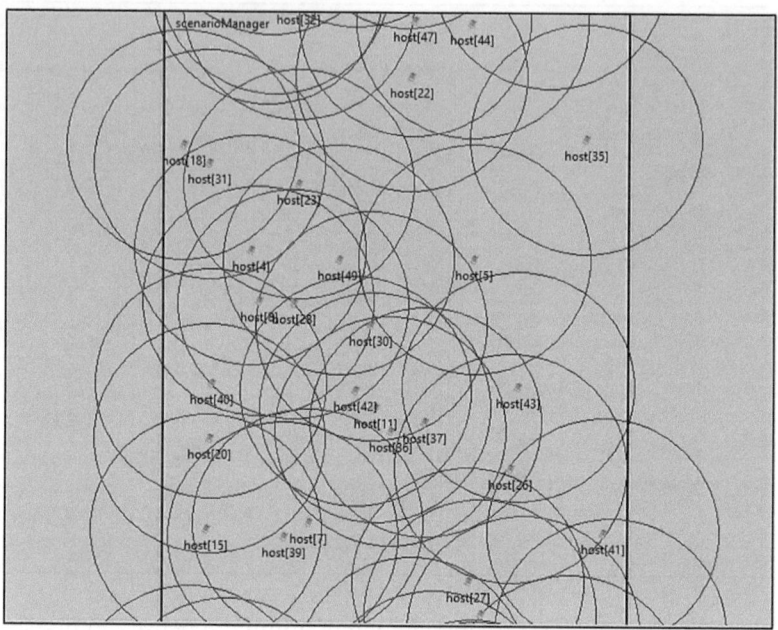

FIGURE 3.3 Omnet++ representation of minimum density nodes moving around.

experimental networks (testbeds), or software-based simulators. Analytical modeling involves making certain simplifications and predictions about network performance. However, overly simplified models or incorrect predictions can lead to inaccurate results.

Testbeds are typically used to create real-world application scenarios on physical hardware. Since these experiments utilize actual equipment, the results tend to be highly accurate. However, due to the cost of real hardware, testbeds are often limited to small-scale applications with a relatively small number of nodes.

For cost-effective experiments, simulations offer a compelling alternative as they can be conducted without the need for physical hardware. Simulations provide flexibility in modeling MANETs with large queue sizes, significant bandwidth, and a high number of nodes. Additionally, simulation results are easier to analyze, as

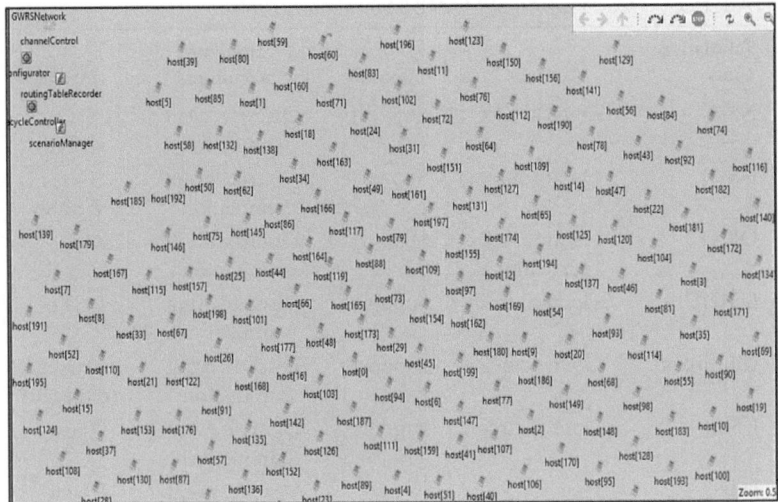

FIGURE 3.4   Omnet++ representation of maximum density in disaster area.

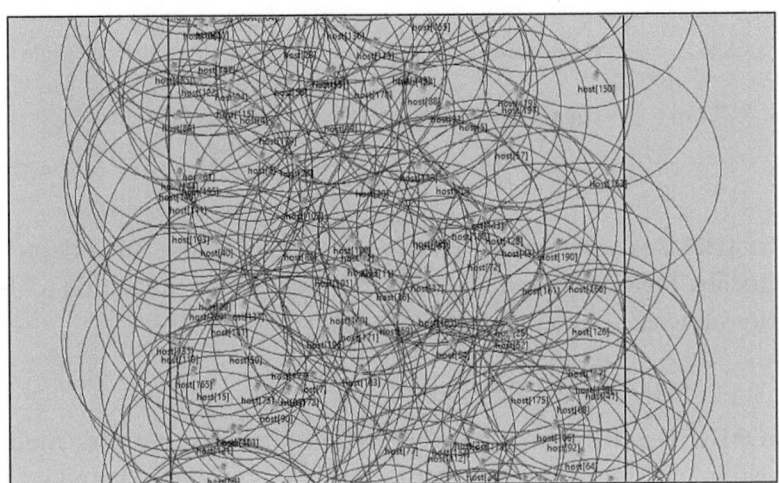

FIGURE 3.5   Omnet++ representation of maximum density nodes moving around.

TABLE 3.1   Comparison of MANET Simulation Tools

Simulation Tools	Type	Mobility	Simulation Technique	Interface
NS-2	Open source	Support	Discrete event simulation	C++ / OTCL
NS-3	Open source	Support	Discrete event simulation	C++ / Python
OPNET	Commercial / academic	Support	Discrete event simulation	C
OMNET++	Open source	Support	Discrete event simulation	C++
GloMoSim	Open source	Support	Discrete event simulation	Parsec (C-based)
J-Sim	Open source	Support	Discrete event simulation	Java
Jane	Free	Native	Discrete event simulation	Java
QualNet	Commercial	Support	Discrete event simulation	Parsec (C-based)
SWANS	Open source	-	Discrete event simulation	Java
GTNets	Open source	No	Discrete event simulation	C++
NAB	Open source	Native	Discrete event simulation	OCaml
NCTUns	Open source	Support	Discrete event simulation	C

critical data points can be readily logged to diagnose network protocols. Table 3.1 presents a list of commonly used simulation tools for network simulation tasks [82, 83].

3.4.2 Selecting the Best Tool

OMNET++ was chosen as the preferred tool for network modeling and simulation due to its widespread availability and reputation. This simulation tool is well-crafted, developed in C++, and boasts open-source accessibility. It is distinguished by its robust GUI support, which significantly simplifies the tracing and debugging processes in comparison to other simulation tools [84]. OMNET++ empowers users to graphically design and construct network simulation scenarios, providing an accurate representation of the simulation during execution. Scenario topologies can be generated as network description (NED) files, and the tool facilitates hierarchical modeling, allowing for zooming into component-level details and displaying the state of each component during the simulation to observe data flow and node interactions.

In OMNET++, the fundamental entity is a module, and each module possesses its own behavior and can be configured as a submodule. Modules communicate with one another by sending and receiving messages through connections. OMNET++ has the capacity to simulate complex IT systems, such as queuing networks and hardware architectures. Additionally, it features an NET extension framework that supports wireless and mobile network simulations. Numerous network researchers have leveraged OMNET++ for the simulation and performance assessment of MANETs [82, 83, 9].

3.5 PERFORMANCE MEASUREMENT METRICS

Throughput stands as a crucial criterion in the assessment of network performance [85]. To measure the effectiveness of the proposed GWRS, various performance metrics [86], including network packet throughput, packet end-to-end delay, packet delivery rate, and packet loss ratio, were employed.

3.5.1 Average End-to-End Delay

The end-to-end delay refers to the duration it takes for a packet to traverse from its source to its destination. This delay can be influenced by factors, such as node mobility, queuing at nodes due to congestion, and packet retransmission. The average end-to-end delay can be mathematically expressed as follows:

$$\text{mean } p = \frac{\sum_{i=1} P_i}{N}, \sum P_i \leq T \tag{3.1}$$

where P_i represents the transmission time of an individual packet when i ranges from 1, 2, 2 .... $\leq T$, N is the total number of packets that have arrived at the destination node, and T is the simulation time, which is 900 s in our parameter environment.

P is given by the following:

$$P = \Delta t = t_r - t_s$$

where t_r and t_s are packet received time and packet sent time, respectively.

3.5.2 Packet Loss Ratio

The packet loss ratio is the number of dropped packets during packet transmission. The packets can be dropped because of excessive waiting time or if the route is broken. Packet loss ratio can be formulated as follows:

$$L = \left(\frac{\sum S_i - \sum R_i}{\sum S_i} \right) \times 100 \tag{3.2}$$

where L is a packet loss ratio, S_i is the number of packets sent by source nodes, and R_i is the number of packets received by destination nodes when $i = 1, 2 \ .... \leq T$.

3.5.3 Packet Delivery Ratio

The packet loss ratio is a metric that quantifies the proportion of packets that are lost during the transmission process. These losses can occur due to prolonged waiting times or disruptions in the communication route. The formula for calculating the packet loss ratio is as follows:

$$D = \left(\frac{\sum R_i}{\sum S_i} \right) \times 100 \tag{3.3}$$

where D is the packet delivery ratio, R_i is the number of packets received by destination nodes, and S_i is the number of packets sent by source nodes when $i = 1, 2 \ldots \leq T$.

3.5.4 Throughput

Throughput is equal to the total number of successfully delivered messages divided by the simulation time:

$$\text{mean}\, q = \frac{\sum R_i}{T} \tag{3.4}$$

where q is a packet throughput.

3.6 SIMULATED SCENARIOS

In this simulation, we used two scenarios to represent disaster recovery areas.

3.6.1 Scenario 1

In the first scenario, we employed a set of general environmental parameters. The designated area spanned dimensions of 1,200 meters by 800 meters, accommodating 100 MNs distributed throughout the region. These MNs in a MANET establish interconnections via multi-hop communication paths or radio links, allowing each MN to move randomly at various speeds and in any direction. Specifically, we configured this simulation using IEEE 802.11b (2.4 GHz), with a transmission range of 250 meters for each node. The default propagation model utilized was the free space model, integrated into the INETMANET framework of OMNET++. We employed a random waypoint model, setting the mobility speeds at 2 Mbps with a data rate of 2 Mbps. Regarding traffic type selection for the disaster area scenario, while conventional wisdom suggests encouraging people to use text messaging to alleviate network congestion, modern smartphone technology often leads individuals to share real-time video footage of disaster situations with their family and friends. Consequently,

TABLE 3.2   Scenario 1 Simulation Parameter

Parameter	Value
Simulation area (m²)	1,200 × 800
Simulation time (s)	900
Mobility model	Random waypoint
Mobile node placement	Random
Pause time (s)	0–2
Gateway	3
Traffic type	CBR
Wireless MAC interface	IEEE 802.11b
Radio propagation model	Free-space model
Packet size	512 bytes
Node speed (Mbps)	2
Transmission range (m)	250
Number of nodes	100

we incorporated CBR (Constant Bit Rate) traffic as a widely used model for simulating data traffic in the scenario.

In this specific simulation, we configured 100 nodes, designating nodes 8, 15, and 49 as gateways. These three nodes were assumed to possess wireless Internet coverage, making them gateways for all MANET nodes without Internet access. The simulation duration was set at 900 seconds. The gateways first initialized their current positions and then determined their neighbors. Subsequently, nodes at each level identified their neighbors to establish the shortest route to the gateway. An overview of the simulation environment is presented in Table 3.2.

3.6.2 Scenario 2

The scope of simulation in Scenario 1 was broadened to mirror the environmental conditions of a real-world disaster scenario. The second simulation was designed to replicate a disaster-stricken area in Loja City, Ecuador [72], utilizing the same simulation tools. In this case, the environmental dimensions were adjusted to 1,000 meters by 2,000 meters. As detailed in Table 3.3, certain

TABLE 3.3   Additional Simulation Parameter

Parameter	Value
Simulation area (m²)	1,000 × 2,000
Number of nodes	50, 97, 100, 120, 160, 200
Number of gateway	1, 3, 6, 10, 15, 20
Nodes speed (mps)	Uniform (0–2)

FIGURE 3.6   Loja City on Google maps.

parameters underwent modification to showcase distinct out-comes resulting from increased MN density and the number of gateways in the disaster-stricken region. However, several other parameters were retained at their previous settings.

Considering the geographical area of Loja City, as illustrated in Figures 3.6 and 3.7, the node densities within the simulation area ranged approximately between 97, 100, 120, and 160. Additionally, the number of nodes was configured to vary from a minimum of 50 to a maximum of 200 for this simulation. In the initial scenario, only 3 gateways were utilized with 100 MNs. Conversely, the second scenario featured an escalating number of gateways, ranging from a minimum of 1 to a maximum of 20.

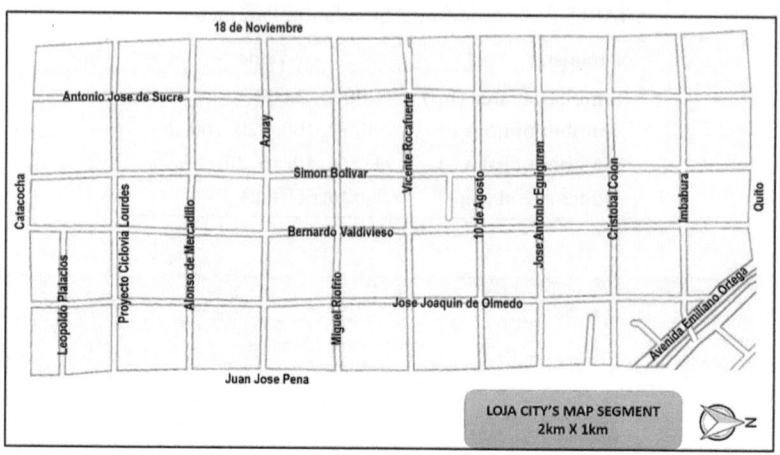

FIGURE 3.7   Loja City map segment [72].

In each simulation, the number of gateways was systematically altered to explore the relationship between node density and the quantity of gateways. Furthermore, this scenario incorporated node mobility within the simulation area to emulate a realistic disaster-stricken environment.

3.7 SUMMARY

In this chapter, we introduced an efficient routing selection scheme designed to alleviate network congestion within disaster recovery areas. We utilized a realistic disaster recovery scenario to assess and compare the performance of our proposed scheme against the AODV and DSDV routing schemes. These evaluations were conducted using OMNET++, a preferred tool for MANET environments.

The primary focus of our research was to streamline the routing selection process within a mobile environment, aiming to simplify the original routing table's complexity. The key importance of our proposed scheme lies in its ability to mitigate network congestion, leading to enhanced packet transmission and overall MANET performance.

3.8 KEY TERMS

AODV, MANET, Reactive Routing, RREQ, Broadcast Storm, Routing Overhead, Disaster Recovery Scenarios, Routing Selection Scheme, Gateways, Network Congestion, Packet Delay, Packet Loss Ratio, Packet Delivery Rate, Throughput, OMNET++, Node Density, Routing Table, Mobility Model, CBR, Scalability.

3.9 REVIEW QUESTIONS

1. What is the primary advantage of the AODV routing protocol in MANETs, and what is its main drawback?

2. How does the proposed routing selection scheme aim to address the challenges of routing in disaster recovery areas?

3. Explain the significance of network congestion and its impact on disaster recovery communication.

4. What performance metrics are commonly used to evaluate the efficiency of routing schemes in MANETs, and how are they calculated?

5. Why is OMNET++ chosen as the simulation tool for modeling MANETs, and what are its key features?

3.10 MINI PROJECTS

These mini projects offer practical experience in MANET simulation, routing protocol implementation, and performance analysis. These projects allow you to apply the knowledge acquired throughout this chapter to real-world scenarios.

1. Create a simulated environment in OMNET++ and implement the AODV routing protocol. Evaluate its performance metrics in various scenarios, such as different node densities and mobility patterns.

2. Design a comprehensive disaster recovery simulation using OMNET++. Consider factors like varying levels of disaster

severity, real-time traffic generation, and dynamic gateway placement. Assess the effectiveness of different routing protocols in such scenarios.

3. Develop a simulation model to investigate energy-efficient routing in MANETs. Implement mechanisms for conserving node energy while maintaining communication in disaster recovery areas.

4. Evaluate the scalability of routing protocols in MANETs by gradually increasing the number of nodes and gateways in the simulation environment. Measure how performance metrics change with network size.

CHAPTER 4

Simulation Results

LEARNING OBJECTIVES

After reading and completing this chapter, you will be able to:

- Discuss the challenges of communication breakdowns in disaster recovery scenarios and the importance of Mobile Ad Hoc Networks (MANETs).

- Discuss the impact of network congestion in MANETs during disaster recovery and the need for efficient gateway selection schemes.

- Discuss the factors influencing mobility patterns in disaster recovery scenarios, including node speed, direction, position, and pause time.

- Analyze the performance metrics of MANETs in disaster recovery, such as throughput, packet delay, packet loss ratio, and packet delivery ratio.

- Evaluate the proposed gateway selection scheme (GWRS) in managing network congestion and enhancing packet transmission efficiency.

DOI: 10.1201/9781032700571-4

4.1 INTRODUCTION

In the context of a disaster recovery scenario, the breakdown of communication infrastructure can result in communication failures. While Mobile Ad Hoc Networks (MANETs) can be deployed for applications like disaster recovery, they often face congestion due to a surge in data traffic as victims attempt to contact their families and friends. Typically, in MANETs, nodes send packets to the nearest gateway to establish external network connections, irrespective of the load on that gateway. To address this issue, a gateway selection scheme has been introduced to efficiently manage network traffic load, distributing tasks evenly among all gateways.

Alternative methods may prioritize routes with fewer hop nodes as the shortest paths, but this can lead to bottlenecks that hinder overall network performance. MANETs, characterized by the mobility of nodes and wireless connections, are highly impacted by node mobility, which can cause disruptions during data transmission due to the dynamic nature of node connections.

The primary focus of this research has been to simplify the routing selection process in mobile environments, thereby reducing the complexity of the original routing table. The proposed scheme's significance lies in its ability to alleviate network congestion, consequently improving packet transmission and overall MANET performance.

To assess the performance of the proposed GWRS scheme and understand the effects of MANET node mobility in disaster recovery areas, a comprehensive performance analysis was conducted. The objective was to identify a suitable mobility pattern for simulation that could accurately mimic the movement of real victims in disaster recovery scenarios. Considerations included node speed, direction, position, and how nodes moved within the area. Despite the prevalence of the random waypoint mobility model in MANETs, it was chosen for its ability to represent nodes' random speed and direction changes within the designated area, making it a suitable choice for this study. Additionally, this model was selected for its simplicity and widespread usage. A predefined pause time was incorporated into the model, representing obstacles encountered by victims in the disaster recovery area. When nodes encountered obstacles, they would come to a halt for a specified duration, ranging from 1 to 300 seconds, depending on the obstacle's complexity.

All nodes within the simulation area were mobile and represented three groups of people within the disaster recovery area: individuals staying in one place, pedestrians with speeds ranging from 1 m/s to 2 m/s, and individuals in vehicles with speeds ranging from 5 m/s to 32 m/s. The simulation scenario was based on a natural disaster in the Loja City area, which resulted in the failure of the public communication network and involved a population density ranging from 50 to 200 people.

The impact of node mobility in disaster recovery was thoroughly investigated using simulations with various mobility speeds. Network performance was assessed in terms of throughput, average delay, packet drop ratio, and sent packet rate. The obtained results were compared to previous work that evaluated Ad Hoc On-Demand Distance Vector (AODV), Destination-Sequenced Distance-Vector (DSDV), Dynamic Source Routing (DSR), and Optimized Link State Routing Protocol (OLSR) protocols under varying conditions of mobility, speed, and node density. DSR demonstrated the best performance among the four protocols at speeds ranging from 20 m/s to the maximum, achieving throughputs between 120 and 140 Kbps. AODV displayed a peak throughput of 167.5 Kbps with 80 nodes. However, these schemes struggled to maintain performance when the network load exceeded 80 nodes. This highlights the significance of the proposed GWRS scheme as a solution for effectively managing large networks during disaster scenarios.

4.2 DELAY PERFORMANCE

4.2.1 Gateway Average Delay

Average delay is a metric employed to determine the average time it takes for packet data from all nodes within a MANET to

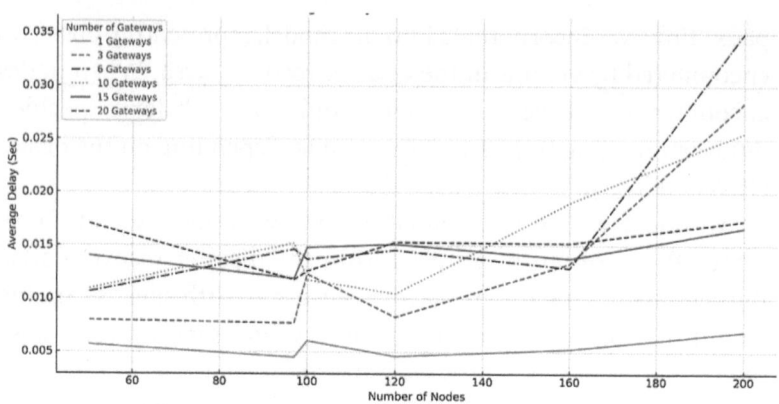

FIGURE 4.1   Gateway average delay versus number of nodes.

reach their final destination. Figure 4.1 provides insights into the load distribution among gateways 1, 15, and 20, which remained relatively consistent even with an increasing number of nodes. However, for gateways 3, 6, and 10, the graph illustrates a sudden surge in average delay when the number of nodes reached 200. Notably, when there were 160 Mobile Nodes (MNs), the network with 10 gateways showed a slightly higher average delay compared to networks with 1, 3, 6, 15, and 20 gateways. While the configuration with a single gateway exhibited the lowest delay across varying node counts, it also highlighted the maximum capacity that a single gateway could effectively handle.

4.2.2 Routing End-to-End Delay for 20 Connections

End-to-end delay signifies the duration it takes for packets to traverse from their source to their destination. This encompasses delays introduced during route discovery, buffer queuing due to congestion, and packet retransmission. Figure 4.2 provides an overview of the outcomes obtained from the Loja City area simulation, featuring node densities ranging from 50 to 200, with 20 randomly established connections. The bar chart illustrates that the proposed scheme exhibited a gradual increase in end-to-end

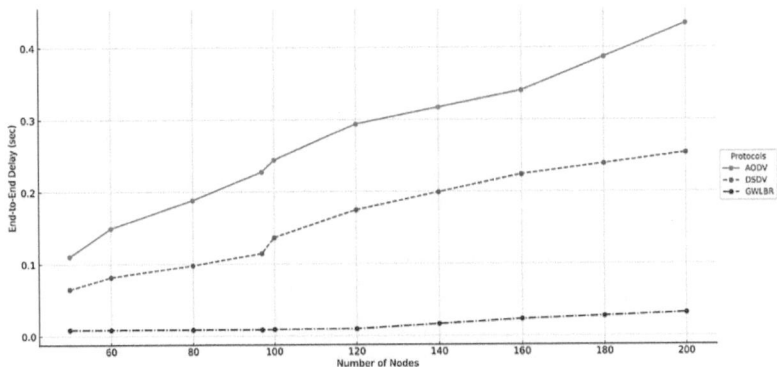

FIGURE 4.2   Routing end-to-end delay for 20 connections.

delay as the number of nodes grew. Nevertheless, it's worth noting that the proposed scheme maintained a lower delay compared to AODV and DSDV.

4.2.3 Routing End-to-End Delay for 40 Connections

Additional simulations were conducted with 40 connections, and the results are depicted in Figure 4.3. In this context, the proposed scheme exhibited a more modest increase in delay compared to DSDV. However, it's noteworthy that DSDV schemes outperformed AODV, with AODV displaying the highest packet delay when the node count was 50. This delay further escalated as the number of nodes reached 200.

4.2.4 Mobility Average Delay

Figure 4.4 illustrates a significant change on average delay as the number of nodes increased. The highest delay was observed at the maximum mobility speed and the highest node count in the disaster recovery area. Given the mobility of nodes in this scenario, when a source node attempted to transmit a packet to its next neighbor, the neighboring node could suddenly move out of the coverage range of the source node. In scenarios involving static and slow movement speeds, which represent people

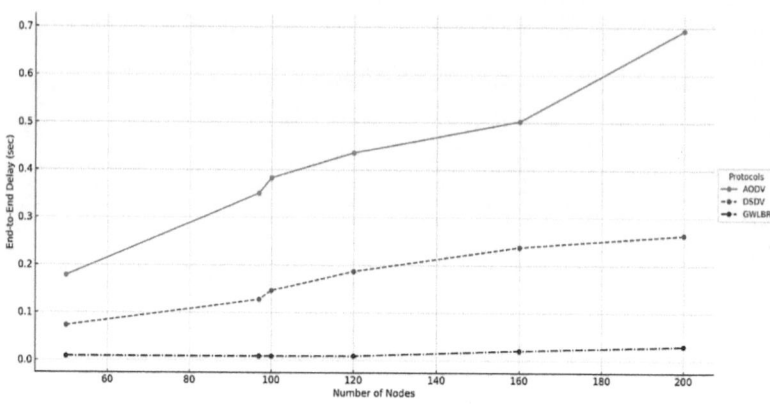

FIGURE 4.3   End-to-end delay for 40 connections.

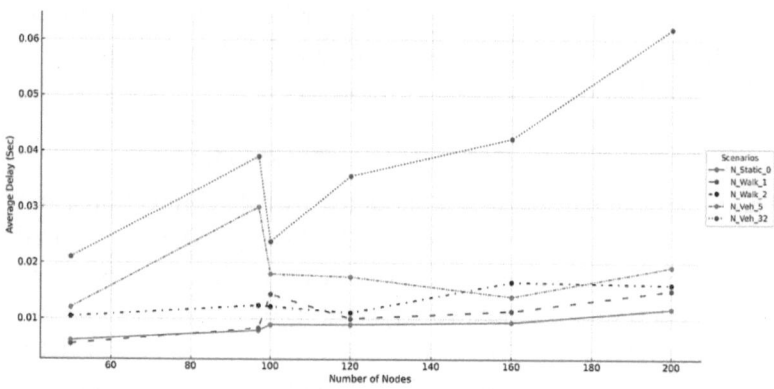

FIGURE 4.4   Average packet delay for mobility speed.

walking, the proposed GWRS exhibited a low initial delay that gradually increased with higher node movement speeds. The density of nodes in the disaster area, coupled with high-speed movement (32 m/s), contributed to the elevated average delay.

4.2.5 Pause time Average Delay

The analysis reveals distinct patterns in packet delay based on node mobility. Nodes representing individuals walking exhibited lower packet delays, particularly with various pause times,

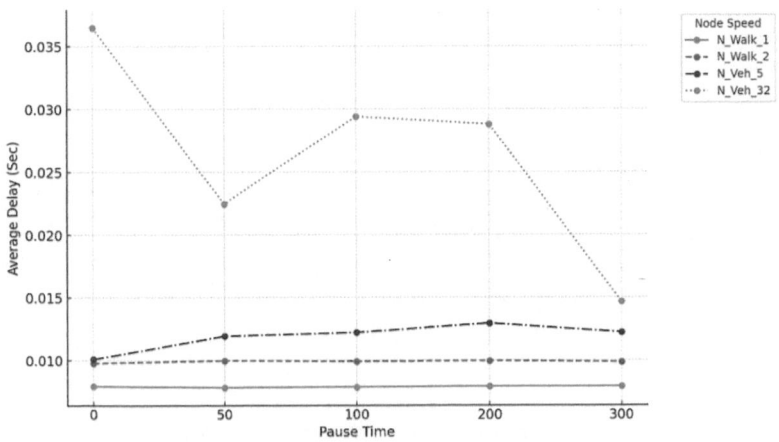

FIGURE 4.5   Packet delay versus pause time.

indicating better performance in disaster recovery scenarios for slowly moving nodes. Packet delay for nodes simulating vehicular movement at a minimum speed of 5 m/s showed a slight increase, with no significant disparity compared to nodes mimicking walking individuals with a maximum speed of 2 m/s. However, individuals in vehicles traveling at maximum speed experienced the highest delays. Interestingly, these vehicular groups experienced reduced packet delay when the pause time was extended to a maximum of 300 s. Notably, there was only a marginal 0.002 s difference in packet delay for nodes with a mobility speed of 5 m/s. In summary, Figure 4.5 highlights that as node mobility approached maximum speed, packet delay increased when nodes moved continuously, while longer pause times proved advantageous for nodes operating at high velocities.

4.3 PACKET LOSS RATIO

4.3.1 Gateway Packet Loss Ratio

Figure 4.6 illustrates that a single gateway configuration in the network struggled to handle the traffic flow to another network,

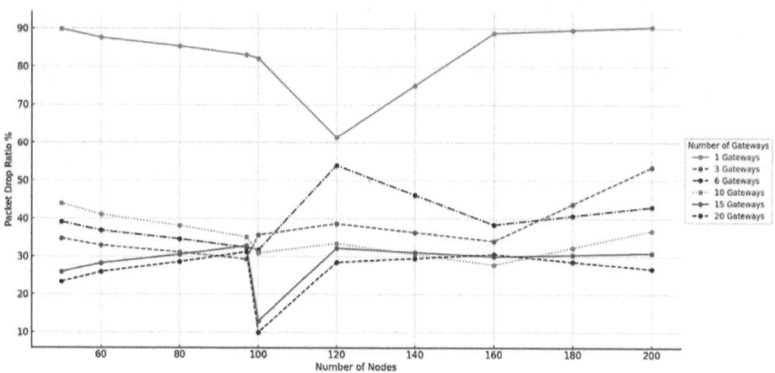

FIGURE 4.6   Packet drops ratio versus number of nodes.

resulting in a minimum packet drop exceeding 60%, marking the highest packet drop rate observed. Notably, the graph highlights the unreliability of the gateway selection scheme when only one gateway is employed. This issue necessitates the implementation of a routing scheme as a potential solution. Examining the results, it becomes evident that the proposed gateway selection scheme effectively managed traffic flow from the disaster recovery area, maintaining a packet drop rate below 45%, even when dealing with the maximum number of nodes. However, exceptions occurred in the cases of 3 and 6 gateways when the number of MNs reached 120 and 200, respectively.

4.3.2 Routing Packet Loss Ratio for 20 Connections

In the context of disaster recovery, the mobility of nodes reflects real-world scenarios involving individuals equipped with mobile devices. This mobility leads to rapid changes in network topology. Figure 4.7 illustrates that when nodes remain static, the end-to-end delay experiences a slight increase. The bar chart provides valuable insights. First, at a node density of 50, the packet loss ratio exhibited a similar pattern across the three schemes. Second, as the number of nodes increased, the charts for AODV and DSDV displayed comparable trends.

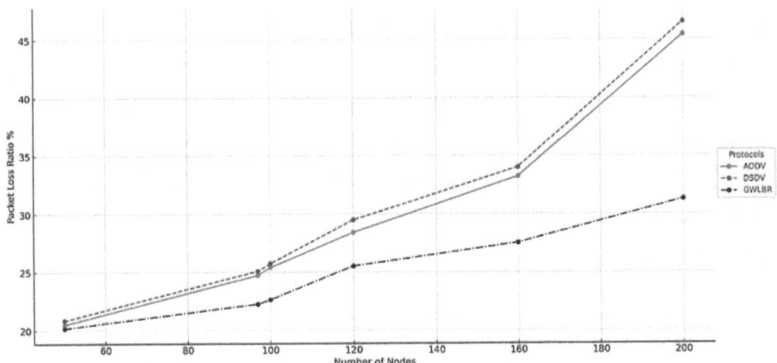

FIGURE 4.7   Packet loss ratio for 20 connections.

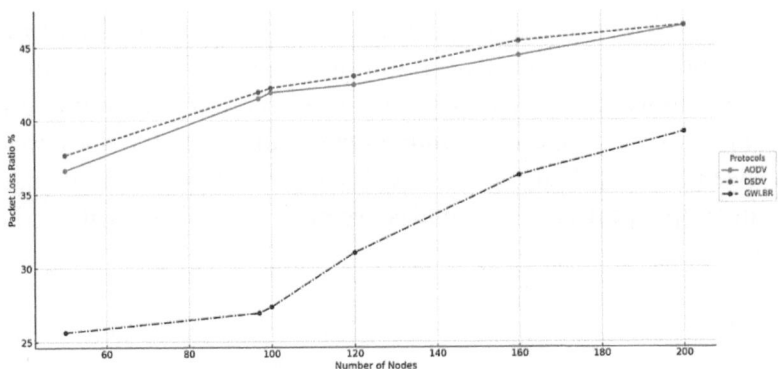

FIGURE 4.8   Packet loss ratio.

However, despite the cumulative nature of these three protocols, our proposed scheme consistently demonstrated a lower packet loss ratio. At a node density of 200 nodes, the proposed scheme reached 31%, whereas AODV and DSDV recorded higher rates of 45% and 46%, respectively.

4.3.3 Routing Packet Loss Ratio for 40 Connections

Figure 4.8 displays the outcomes for 40 connections with 50 nodes. The packet loss ratios were observed to be 36% for AODV, 37%

for DSDV, and 25% for the proposed scheme, with the proposed scheme demonstrating the lowest packet loss ratio. As the number of nodes doubled, the loss ratios for AODV and DSDV remained similar, while the proposed scheme's loss ratio increased to 27%. With a further doubling of nodes, the loss ratio for the proposed scheme gradually increased to 39%, still maintaining the lowest packet loss ratio.

4.3.4 Mobility Packet Lost Ratio

The performance analysis was extended to include the evaluation of the packet drop ratio. In Figure 4.9, the impact of static mobility and a high mobility speed of 32 m/s (representative of individuals in vehicles in the disaster recovery area) on the packet drop ratio can be observed. Interestingly, a similar pattern emerged as the number of MNs increased, resulting in an elevated packet drop ratio. However, this pattern was not observed for the minimum and maximum speeds of individuals walking or for the lowest speed of individuals in vehicles. The bar chart depicts a fluctuating graph, particularly as nodes began to move slowly within the area.

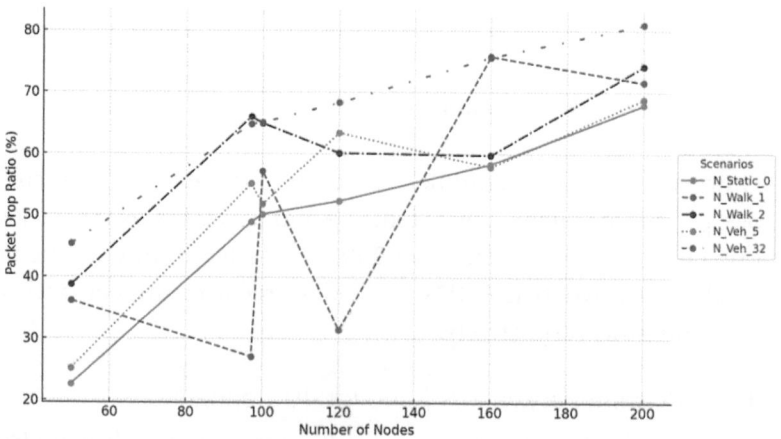

FIGURE 4.9 Packet drops ratio.

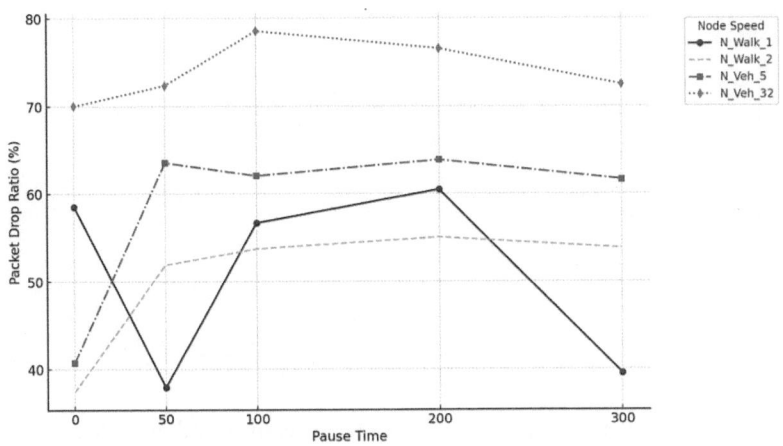

FIGURE 4.10   Packet lost ratio versus pause time.

4.3.5 Pause Time Packet Lost Ratio

Figure 4.10 illustrates that node operating at the maximum speed experienced a consistently high packet drop rate regardless of the pause time. Conversely, nodes traveling at the minimum speed demonstrated the best performance, with only a 39% packet drop rate observed at the maximum pause time of 300 s. Notably, when there was no pause time, indicating continuous movement of nodes, those representing people walking at the maximum speed performed best, with only a 37% packet drop ratio. This suggests that a node speed of 2 m/s without pause time can reduce packet drop rates. However, considering that obstacles are typically encountered in disaster recovery areas, causing nodes to pause before finding alternative routes, the best performance can be achieved when nodes move slowly to minimize packet drops in the presence of obstacles.

4.4 PACKET DELIVERY RATIO

4.4.1 Gateway Packet Delivery Ratio

Further analysis, as depicted in Figure 4.11, confirmed the expected trends in packet drop and packet delivery performance: the single

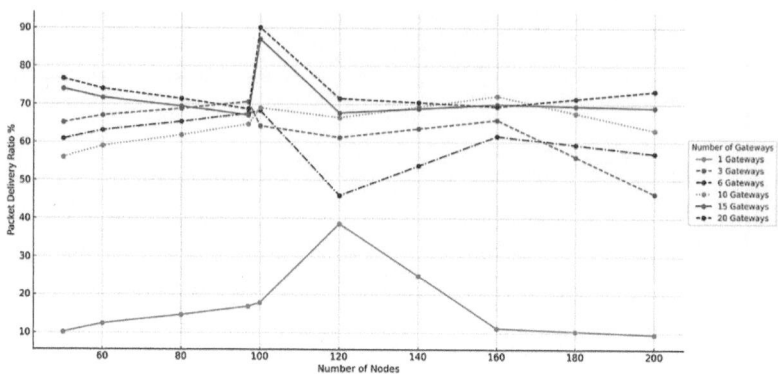

FIGURE 4.11   Gateway packet delivery ratio versus number of nodes.

gateway exhibited the lowest packet delivery ratio. However, when the proposed selection scheme was applied to gateways, packet delivery became considerably more efficient, regardless of the number of gateways involved. The packet delivery ratio reached over 70% for 3 gateways and peaked at 90% when 20 gateways were integrated into the disaster recovery network. On average, packet delivery from the minimum to 200 MNs in the disaster area, with three gateways, reached 62%. This figure slightly increased to 65% for 10 gateways and significantly improved to 75% for 20 gateways. These results underscore the significance of developing an efficient gateway selection scheme to ensure higher packet delivery rates.

4.4.2 Routing Packet Delivery Ratio

Packet delivery refers to the ratio of successfully delivered packets to their intended destination nodes. As illustrated in Figure 4.12, when there were 50 nodes in the disaster area, approximately 80% of the packets reached their destination nodes successfully. However, as the number of nodes increased to 97, the gap in packet loss ratios between DSDV and the proposed scheme narrowed to only 3%. This gap continued to widen as the number of nodes further increased. Notably, packet delivery gradually declined for

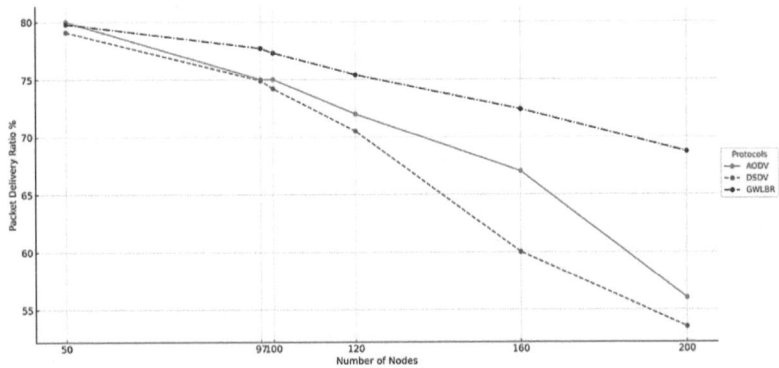

FIGURE 4.12   Routing packet delivery ratio for 20 connections.

the DSDV scheme, making it the least effective option when compared to the other two schemes.

4.4.3 Mobility Packet Delivery Ratio

The implications of these findings become evident when considering the packet delivery ratio results shown in Figure 4.13. When there were 50 nodes in the disaster recovery area and individuals remained stationary, the highest packet delivery ratio was achieved. Conversely, the lowest packet delivery ratio was observed when 200 nodes were present in the area, and the mobility speed was increased to the highest level, representative of extreme vehicle speeds. This decrease in packet delivery can be attributed to the fact that as the number of nodes increases, the GWRS routing scheme must deliver packets to a greater number of destinations, leading to a reduction in overall delivery efficiency. The packet delivery ratios exhibited fluctuations in results for mobility speeds of 1 m/s, 2 m/s, and 5 m/s, with varying node densities, excluding the scenario with 200 nodes. These fluctuations were most pronounced for vehicles traveling at a speed of 5 m/s, followed by individuals walking at a speed of 1 m/s. Interestingly, at a node density of 200, the disparity between these speeds was only 2.7%. In this context, individuals walking at a speed of 2 m/s

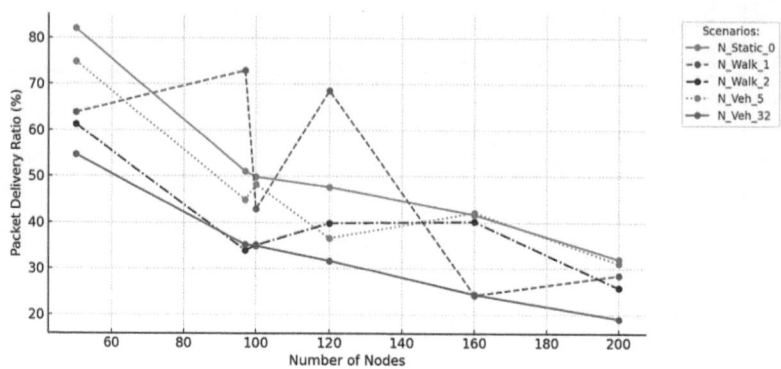

FIGURE 4.13   Mobility packet delivery ratio.

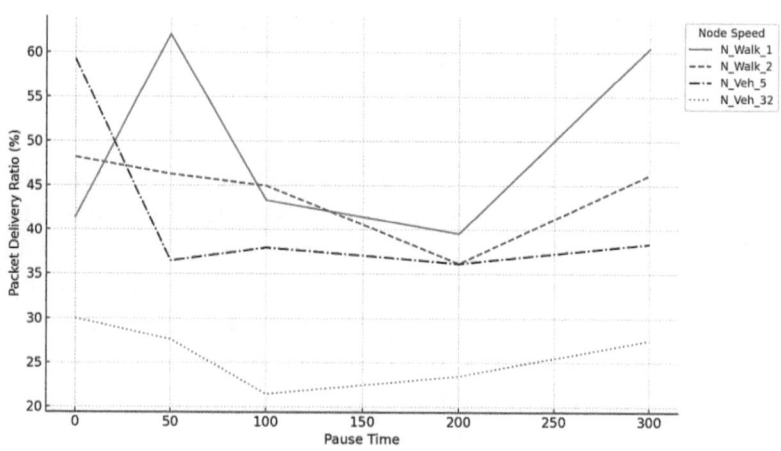

FIGURE 4.14   Packet delivery ratio versus pause time.

initially had the lowest delivery ratio but eventually reached an intermediate position when the node density reached 120 nodes (Figure 4.13).

4.4.4 Pause Time Packet Delivery Ratio

The bar chart presented in Figure 4.14 illustrates the packet delivery ratio, with results indicating that individuals traveling in vehicles at maximum speed experienced lower packet delivery rates. Even with an increased pause time, allowing nodes more time to make optimal routing decisions, the delivery ratio did not surpass 30% in this scenario. However, as the node mobility speed decreased to 5 m/s, packet delivery rates improved, rising from 36% to 59%. In the case of individuals walking, performance exhibited fluctuations as node mobility speed decreased and pause time increased. When individuals walked at maximum speed with no pause time, the packet delivery ratio stood at only 62%. Interestingly, performance gradually declined as pause time increased in this scenario.

4.5 THROUGHPUT

4.5.1 Gateway Throughput

In this simulation, all MNs had access to Information Gateways (IGs) through neighboring nodes. The results depicted in Figure 4.15 demonstrate that different gateways yielded varying outcomes, dependent on the load capacity of the serving gateway. Significant fluctuations were observed when only a single gateway node was available to cater to user demands. This was a consequence of increasing load on a single gateway as the number of MNs grew. With the introduction of three gateways, fluctuations persisted but were somewhat mitigated since the load could

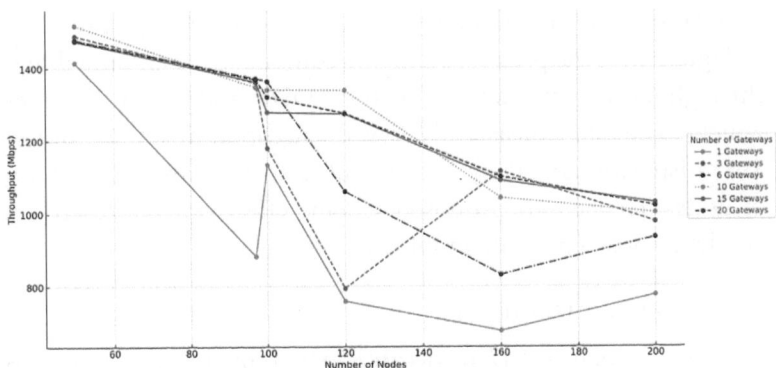

FIGURE 4.15   Gateway throughput versus number of nodes.

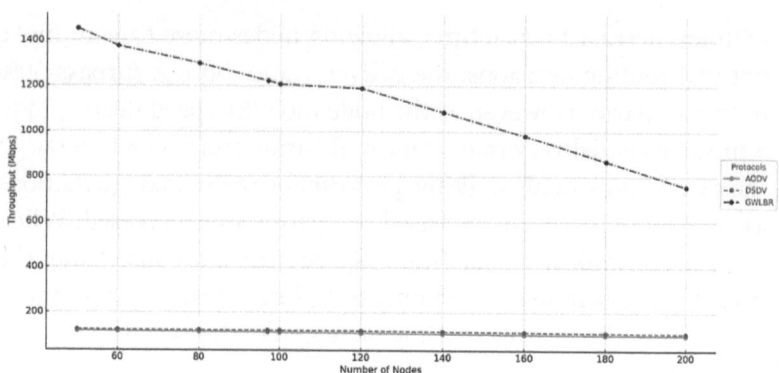

FIGURE 4.16   Routing packet throughput.

be distributed among multiple gateways. Notably, as the number of gateways increased, the network's performance exhibited improved throughput. However, as node density increased, packet throughput experienced a reduction.

4.5.2 Routing Throughput

Our primary objective was to minimize packet loss, recognizing the critical importance of communication during disaster recovery when demand is exceptionally high. Another noteworthy discovery was that our proposed scheme achieved a significant outcome by offering improved basic Internet access to the user population in the recovery area. As evident in the graph presented in Figure 4.16, the proposed scheme consistently maintained high throughput, a stark contrast to the AODV and DSDV schemes, which exhibited consistently low throughputs from the outset. This disparity can likely be attributed to the random mobility of nodes. Hence, our proposed scheme effectively addressed the challenge posed by node mobility.

4.5.3 Mobility Throughput

As shown in Figure 4.17, it was anticipated that in static scenarios, nodes would exhibit the highest average throughput. However,

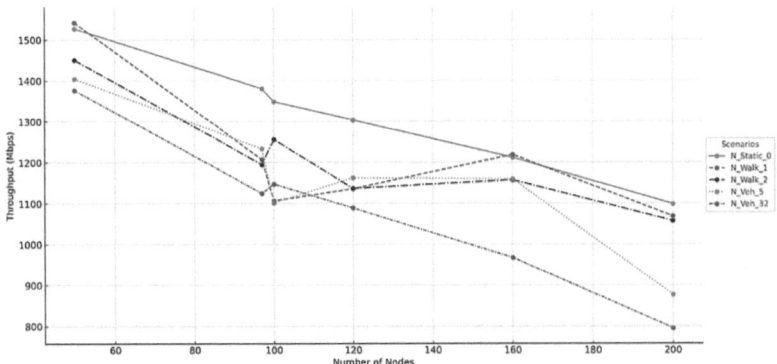

FIGURE 4.17   Throughput of node mobility.

there was a surprising revelation when considering nodes representing people walking in the disaster area, with velocities ranging from 1 to 2 m/s. Even as the node density increased from 50 to 200, the throughput consistently remained above 1 Mbps. In the case of nodes in vehicles traveling at maximum speeds, the throughput dropped below 1 Mbps beyond a certain node density threshold but still maintained levels above 700 Kbps. AODV performed admirably, achieving a peak throughput of 167.5 Kbps with 80 nodes. However, these schemes struggled to maintain their performance when subjected to network loads exceeding 80. This underscores a significant advantage of the proposed GWRS scheme in effectively managing large networks during disaster scenarios.

4.5.4 Pause Time Throughput

This book employed a MANET within a disaster recovery area as a case study to assess the performance of the GWRS scheme. The concept of pause time was introduced, wherein nodes would pause at specific locations before determining their subsequent direction and then proceed toward their ultimate destination. This iterative process was reiterated throughout the study.

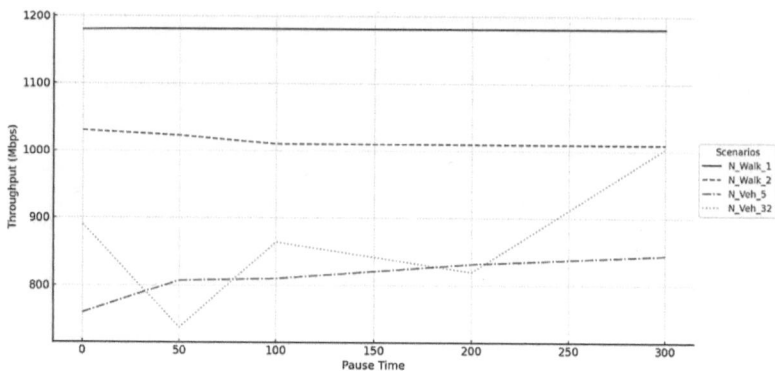

FIGURE 4.18   Throughput versus pause time.

The bar chart in Figure 4.18 depicts the throughput performance of two distinct groups: walking individuals represented by the pyramid shape, and people in vehicles denoted by the rectangular box. Notably, walking individuals consistently exhibited higher throughput across various pause time variations. These findings underscore that reducing node mobility speed led to a significant increase in throughput. Conversely, as mobility speeds escalated, throughput performance witnessed a decline. In scenarios with heightened mobility, the GWRS scheme encountered challenges in identifying the next hop due to the fast movement of all nodes within the area and GWRS continued to demonstrate superior performance.

4.6 SUMMARY

Our simulation results clearly demonstrate that the proposed scheme effectively reduces network congestion, leading to improvements in throughput, packet delay, and packet loss within MANET performance. The efficient distribution of the load among gateways highlights the ability of our proposed scheme to maximize packet throughput in MANETs. Furthermore, this approach substantially alleviates congestion at each gateway, ultimately enhancing MANET performance by increasing the number of successfully transmitted packets. This chapter has extensively discussed the performance enhancement achieved through the routing selection scheme in MANETs deployed in disaster recovery scenarios.

In disaster recovery areas, obstacles are a common occurrence, often causing nodes to pause at specific locations before continuing their search for alternative routes. It can be inferred that the optimal strategy for maintaining performance in the face of obstacles within a disaster recovery area is to move slowly, thus minimizing packet loss. When node mobility reaches its maximum speed, packet delay tends to increase if nodes do not pause. A longer pause time can greatly benefit nodes with high velocities.

The influence of node mobility on the performance of GWRS schemes in MANETs during disaster recovery is a consistent finding across all the results obtained in this simulation. The implications of these findings for network planning and deployment will be discussed in the next chapter.

4.7 KEY TERMS

MANETs, Disaster Recovery, Network Congestion, Gateway Routing Selection Scheme (GWRS), Node Mobility, Packet Delivery Ratio, Packet Loss Ratio, Packet Delay, Throughput, Routing Selection Scheme, Pause Time, End-to-End Delay, Mobility Speed, Network Load, Node Density.

4.8 REVIEW QUESTIONS

1. What challenges do MANETs address in disaster recovery scenarios, and why does network congestion become a significant issue?

2. How does the proposed gateway selection scheme aim to alleviate network congestion and improve packet transmission efficiency?

3. What factors contribute to node mobility in disaster recovery areas, and why is it important to consider these factors in network simulations?

4. Explain the performance metrics used to evaluate MANETs in disaster recovery, including throughput, packet delay, packet loss ratio, and packet delivery ratio.

5. How does the GWRS scheme impact the distribution of network traffic load, and what role does it play in ensuring efficient packet delivery?

4.9 MINI PROJECTS

1. Develop a simulation or small-scale implementation of the gateway selection scheme (GWRS) in a MANET environment to observe its impact on network congestion and packet delivery.

2. Conduct a study on real-world mobility patterns during disaster recovery scenarios, collect data, and analyze it to better understand node movement and its implications on network performance.

3. Compare the performance of various routing protocols (e.g., AODV, DSDV) with the proposed GWRS scheme under different scenarios (e.g., node density, mobility speed) using network simulation tools.

4. Investigate the optimal pause time duration for nodes in disaster recovery areas to minimize packet loss and delay and propose a scheme to dynamically adjust pause times based on network conditions.

5. Create a comprehensive disaster recovery simulation incorporating node mobility, network congestion, and the GWRS scheme to evaluate its effectiveness in a realistic disaster scenario.

CHAPTER 5

Advances in MANETs for Disaster Recovery

LEARNING OBJECTIVES

After reading and completing this chapter, you will be able to:

- Discuss a typical MANET scenario for disaster recovery, especially historical development, and practical applications.

- Explore potential future trends in MANET technology such as integration with emerging technologies like 5G, Wi-Fi, and the Internet of Things (IoT).

- Comprehend the significance of advanced routing protocols to enhance the performance of MANETs.

- List and discuss two issues/challenges in implementing MANETs in disaster recovery scenarios.

5.1 INTRODUCTION

In this section, we delve into the historical development and progression of Mobile Ad Hoc Network (MANET) and how MANET technologies have developed and changed over time. This

DOI: 10.1201/9781032700571-5

involves a detailed look at the initial inception, key technological advancements, and recent innovations in the field. We also analyze how the evolutionary changes have influenced their application in disaster recovery scenarios. After taking into consideration the historical progression and recent advancements in MANET technologies, we then explore potential future trends and improvements. This includes a discussion on the integration of MANETs with other emerging technologies and the potential for these integrations to further enhance the effectiveness of MANETs in disaster recovery scenarios.

In the forthcoming section of this comprehensive study, we are going to embark on an intricate journey that traverses the historical development and progression of MANET technologies. We delve into the roots of this fascinating technology, starting from its initial inception, moving on to the key technological advancements that have shaped its trajectory, and finally arriving at the most recent innovations that are pushing the boundaries in the field.

As part of this exploration, we also dedicate a significant amount of attention to analyzing how these evolutionary changes, both large and small, have influenced and continue to influence the application of MANET technologies in disaster recovery scenarios. This includes both the successes and challenges that have arisen in the practical implementation of these technologies in real-world situations.

Having thoroughly examined the historical progression and recent advancements in MANET technologies, we then turn our sights to the horizon. Our exploration continues with a detailed forecast on the potential future trends and improvements that could shape the next chapter in the story of MANETs. This part of our discussion will not only be limited to the evolution of MANETs themselves but will also include a deep dive into the integration of MANETs with other emerging technologies that make waves in the tech world.

FIGURE 5.1   MANET setup in a typical disaster area.

Figure 5.1 illustrates a MANET setup in a disaster-affected area. It showcases various elements such as handheld devices, emergency response vehicles, and temporary command centers, all interconnected to form the network essential for communication and coordination in such critical situations. This visualization can help in understanding how a MANET functions during emergency responses.

We will explore the potential for these integrations to further enhance the effectiveness of MANETs in disaster recovery scenarios, including the potential benefits and challenges that could arise from such advancements. This exploration is guided by current research and thought leaders in the field, providing a well-rounded and forward-thinking perspective on the future of MANET technologies.

5.2 ACCESSIBILITY AND INCLUSIVITY IN NETWORK DESIGN

The urgency of the situation demands that MANET interfaces be simple enough for a layperson to understand, yet robust enough to handle the complexities of disaster communication. This section will delve into design principles that prioritize ease of use and minimize the cognitive load on users, ensuring that the technology serves as a helpful tool rather than an additional challenge during a crisis.

The illustration in Figure 5.2 depicts a disaster scenario simulation using MANET technology. It shows a training field where emergency responders are actively engaged in a simulation, utilizing MANET devices. The scene includes temporary shelters, emergency vehicles, and personnel using handheld devices and laptops for communication. A command center is also shown, with screens displaying the MANET network and teams coordinating the simulation. This visual effectively conveys the organized and dynamic nature of disaster management drills and the crucial role of MANET technology in such training exercises.

5.2.1 Evolution of MANET Technologies

The evolution of MANETs stands as a testament to the remarkable strides made in the field of wireless communications. This progression is particularly salient in the integration of emerging wireless technologies, such as 5G and 6G. These newer generations of wireless technology herald a paradigm shift in MANET performance, offering higher data rates, lower latency, and greater reliability. For instance, 5G networks, with their enhanced throughput and reduced latency, have the potential to significantly augment the operational efficiency of MANETs. They allow for rapid deployment in disaster-stricken areas, enabling effective

FIGURE 5.2   Disaster scenario simulation using MANET technology.

communication channels crucial for rescue and recovery operations. Such integrations are actively being researched, with pilot projects exploring the synergy between 5G's advanced features and the flexible, dynamic nature of MANETs.

Moreover, the concept of energy-efficient networking within MANETs is gaining momentum, particularly through innovations in energy harvesting and power-efficient routing protocols. Techniques that harness solar and kinetic energy can extend the operational life of MANET devices, a critical factor in scenarios where traditional power sources are compromised. For example, solar panels integrated into communication devices can harvest energy, ensuring sustained network availability during prolonged recovery efforts. Power-efficient routing protocols are also essential in this regard, as they aim to minimize energy consumption by intelligently managing the data transmission routes and device operation modes, thereby preserving the network's longevity without sacrificing performance.

The confluence of these two domains, advanced wireless technologies and energy efficiency, presents an exciting frontier for MANET development. By addressing the challenges of integrating high-speed wireless technology and devising innovative energy conservation strategies, researchers and practitioners are setting the stage for more resilient and capable MANETs. These networks are not only envisioned to provide robust communication backbones in the wake of disasters but also to adapt proactively to the demands of the dynamic environments they operate in.

As MANETs continue to evolve, they will undoubtedly encapsulate a broader array of technologies and capabilities, further cementing their role as a vital component in disaster recovery and beyond. Each innovation brings us closer to a future where reliable, autonomous, and energy-conscious networks can be deployed anywhere, anytime, fostering a more connected and resilient world.

The journey of MANET technologies began in the late 20th century with simple, decentralized networks designed for military use. Over the decades, MANET technologies have evolved significantly, becoming more robust, efficient, and versatile. Today, they are utilized in numerous applications beyond the military, such as emergency services, mobile commerce, and vehicular networks.

The evolution of MANET technologies is characterized by the continuous improvement of their design and functionality to meet the growing demands of various applications. This includes enhancing network scalability, improving data transmission speed and reliability, and increasing network lifetime and coverage range.

One of the early major advancements in MANET technology was the development of efficient routing protocols. These protocols were designed to manage the dynamic nature of MANETs, where nodes could freely join, leave, and move within the network. This led to the creation of protocols such as DSR (Dynamic Source Routing) and AODV (Ad Hoc On-Demand Distance Vector), which are still widely used today.

In the 21st century, the focus of MANET research shifted towards improving network security and quality of service (QoS). New technologies were developed to deal with issues such as network attacks, data integrity, and privacy. At the same time, advancements in hardware technology led to the creation of more powerful and energy-efficient network nodes.

Today, the evolution of MANET technologies continues, with a focus on integrating them with other emerging technologies like Internet of Things (IoT), cloud computing, and machine learning (ML). These integrations are expected to lead to new applications and improvements in network performance, thereby playing a crucial role in post-disaster recovery and many other fields.

5.2.1.1 Integration with Emerging Wireless Technologies

In the rapidly evolving world of technology, wireless innovations are becoming increasingly prevalent. In the context of this development, the integration of emerging wireless technologies has become a vital area of focus.

This integration includes the adaptation and usage of new wireless technologies, such as 5G and Wi-Fi 6, which promise faster speeds, lower latency, and improved reliability. These advancements are revolutionizing how we connect and communicate, creating new opportunities for businesses and consumers alike.

Moreover, the integration also involves understanding the compatibility of these technologies with existing systems. It is necessary to ensure seamless integration to avoid disruptions and to maximize the benefits of these technologies. This can involve rigorous testing and the development of new protocols and standards.

Furthermore, the integration of these wireless technologies also demands an evaluation of their impact on security. As these technologies evolve, they present new challenges that need to be addressed to ensure the secure transmission of data.

In conclusion, the integration with emerging wireless technologies is a complex yet rewarding endeavor. It has the potential to significantly enhance connectivity and usher in a new era of technological advancements. The integration of emerging wireless technologies with MANETs has opened exciting avenues for research and development. These technologies can potentially augment the capabilities of MANETs, enhancing their efficiency, reliability, and versatility, particularly in disaster recovery scenarios.

One of the promising aspects of this integration is the improvement of network scalability and performance. These emerging wireless technologies can handle a larger number of nodes, thereby allowing for greater network coverage and connectivity. This is particularly crucial in disaster-stricken areas where the infrastructure is damaged and communication is of utmost importance.

Additionally, emerging wireless technologies, such as 5G and Wi-Fi 6, offer higher data transmission speeds. This can significantly improve the efficiency of data sharing within a MANET, which is critical for effective disaster management and recovery. For instance, real-time data about the situation in disaster zones can be disseminated quickly and efficiently, facilitating prompt decision-making and action.

Furthermore, these technologies also promise lower latency. This is vital for applications that require real-time communication, such as remote rescue operations or telemedicine, where even a slight delay can have severe consequences. However, the integration of these technologies with MANETs also presents new challenges. The nature of these technologies necessitates the development of new protocols and algorithms to manage the dynamic and decentralized nature of MANETs effectively. Moreover, issues pertaining to the security and privacy of data transmission also arise. Therefore, while the integration of emerging wireless technologies with MANETs holds great promise for enhancing disaster recovery operations, it is also accompanied by new challenges that need to be addressed. Future research and development efforts need to focus on mitigating these challenges, ensuring the safety, reliability, and efficiency of these integrated networks.

5.2.1.2 Energy-Efficient Networking

Energy-efficient networking focuses on optimizing network resources to reduce energy consumption. This could involve implementing strategies, such as reducing idle power, using lower energy equipment, or optimizing network protocols and architectures. The goal is to maintain high performance and reliability while minimizing the environmental impact. In the context of MANETs, energy efficiency is particularly critical. Given the often limited and valuable power resources in disaster recovery scenarios, the need for energy-efficient networking is paramount. It not only extends the operational lifespan of the network nodes but also enhances the overall effectiveness of communication in the network.

One approach to energy-efficient networking in MANETs is the design of energy-aware routing protocols. These protocols consider the energy consumption of network nodes during route selection, favoring routes that minimize overall energy consumption. This can significantly extend the network's lifetime and improve its durability in disaster recovery scenarios.

Another approach is with energy-efficient hardware. Advancements in hardware technology have led to the development of network devices that consume less power while maintaining high performance. These devices can significantly reduce the energy footprint of the network, making it more sustainable and reliable.

Additionally, techniques such as dynamic power management and sleep scheduling can also contribute to energy efficiency. These techniques involve intelligently managing the power states of network nodes, such as putting nodes into low-power sleep mode when they are not in use. This can significantly reduce idle power consumption and extend the network lifetime.

However, energy-efficient networking also presents its own set of challenges. For instance, there is often a trade-off between energy efficiency and network performance, with attempts to reduce energy consumption potentially impacting the speed and reliability of data transmission. Moreover, implementing energy-efficient strategies can also increase the complexity of network management and operation.

Despite these challenges, the potential benefits of energy-efficient networking, particularly in the context of disaster recovery, make it a vital area of focus. Moving forward, research and development efforts need to continue exploring innovative strategies and solutions to enhance the energy efficiency of MANETs, thereby improving their sustainability and effectiveness in disaster recovery scenarios.

5.2.2 Advanced Routing Protocols

In the evolution of MANETs, the development of advanced routing protocols has been a critical aspect. These protocols are designed to efficiently manage the dynamic and decentralized nature of MANETs, where nodes can join, leave, and move within the network freely.

One of the primary objectives of these advanced routing protocols is to establish reliable and efficient paths between network nodes. This involves handling a range of challenges, including dynamic topology changes, limited bandwidth, and high error rates [87].

Among the advanced routing protocols developed, DSR and AODV are notable. DSR operates by allowing each network node to dynamically discover a source route across multiple network hops to any destination. On the other hand, AODV creates routes between nodes only when the routes are requested by the source nodes, thereby reducing the need to constantly update the network topology. In addition, advanced routing protocols also incorporate strategies to improve network performance and QoS [88]. For example, they may prioritize routes with higher data transmission speeds or lower latency, thereby ensuring efficient data communication within the network.

Furthermore, these protocols also aim to enhance network security. They incorporate mechanisms to authenticate network nodes and protect data integrity, thereby addressing potential security threats within the network. However, the implementation of advanced routing protocols in MANETs is not without challenges. The dynamic and decentralized nature of MANETs necessitates the constant update of routing information, which can lead to high overhead and reduced network performance. Moreover, these protocols must be robust against network attacks and capable of maintaining network operation in the face of node failures.

Despite these challenges, the development and implementation of advanced routing protocols play a critical role in enhancing the performance and reliability of MANETs. As such, they are crucial for the effective use of MANETs in various applications, including disaster recovery. Moving forward, more research is needed to further improve these protocols and develop new ones that can better meet the growing demands of MANET applications. This includes enhancing their scalability, speed, and security, as well as their integration with emerging technologies like IoT and ML.

5.2.2.1 Integration of Advanced Routing Protocols with IoT and ML

As the field of MANETs continues to evolve, the integration of advanced routing protocols with emerging technologies like IoT and ML is becoming increasingly significant. IoT devices generate vast amounts of data that need to be efficiently routed through the network. Advanced routing protocols can facilitate this by optimizing the path selection based on factors like network congestion, energy efficiency, and data transmission speed. Moreover, these protocols can also enhance the reliability of IoT applications by ensuring robust data communication even in the face of network failures or changes in topology.

Meanwhile, ML can bring a new dimension to the operation of advanced routing protocols. By applying ML algorithms, these protocols can learn from past network behaviors to make more accurate and efficient routing decisions. This can significantly enhance network performance, particularly in dynamic and unpredictable network environments.

However, integrating these emerging technologies with advanced routing protocols also presents new challenges. For instance, IoT devices often have limited processing and battery capabilities, which can constrain the operation of these protocols. Meanwhile, the application of ML requires careful consideration of issues like data privacy and algorithmic bias. Therefore, while the integration of IoT and ML with advanced routing protocols holds great potential for enhancing MANET performance, it also necessitates further research and development efforts to address these challenges and uncertainties.

5.2.2.2 ML-Based Routing

ML offers transformative potential for MANETs, particularly in enhancing routing protocols. By incorporating ML models, MANETs can adapt to dynamic conditions, predict network performance issues, and optimize routes in real-time. For instance, reinforcement learning (RL) models have been successfully applied to dynamic route optimization, enabling nodes to learn and select paths based on the experiences of packet transmissions, effectively reducing congestion, and improving the overall network performance.

An example of RL in practice is the Q-routing algorithm, which uses Q-learning to dynamically adjust routing decisions based on historical data and current network conditions. The algorithm allows nodes to "learn" the most efficient route over time, thereby reducing the end-to-end delay. Simulation studies, such as those conducted using NS3 or OPNET, have demonstrated that ML-based routing can lead to significant improvements in packet delivery ratios and throughput compared to traditional protocols.

5.2.2.3 Blockchain for Secure and Decentralized Communication

Blockchain technology offers a robust solution to secure communications in MANETs by enabling a decentralized authentication system. This approach is particularly beneficial in post-disaster scenarios where central authentication systems are likely to be compromised. Blockchain's distributed ledger system ensures that communication between nodes is authenticated and immutable, providing a secure and reliable method for message exchange without the need for a centralized authority.

For example, blockchain can be used to create a secure and trust less communication protocol for MANETs, where each block contains a batch of messages with cryptographic guarantees of integrity and origin. Case studies have explored blockchain for identity verification, secure message dissemination, and resilient data sharing. Theoretical models, like the BlockMANET, propose a blockchain-based architecture for MANETs that secures data exchange by requiring nodes to validate and agree on data before it is added to the network, thus preventing unauthorized access and ensuring data integrity.

In conclusion, ML models and blockchain technology represent cutting-edge advancements in the domain of MANET routing protocols. By leveraging the predictive capabilities of ML and the secure infrastructure of blockchain, MANETs are poised to become more intelligent, efficient, and resilient, particularly in the critical use case of disaster recovery.

5.2.3 Interoperability with Other Networks

A promising strategy for improving disaster communication coverage involves the integration of MANETs with satellite and terrestrial networks. This approach can significantly enhance the range and reliability of the communication system, facilitating efficient data transmission in remote areas or locations with damaged infrastructure.

Interoperability with other networks involves technical complexities. For instance, it would require efficient routing protocols to manage the dynamic nature of these networks. The protocols should be capable of handling a broad spectrum of network conditions, including changes in node mobility, link quality, and network topology.

Furthermore, the integration might also introduce challenges related to network security, latency, and bandwidth usage. Addressing these issues would require innovative solutions, such as the development of robust encryption methods for data security, utilization of efficient data compression algorithms to mitigate bandwidth constraints, and the employment of advanced error correction techniques to ensure reliable data transmission despite latency issues.

Overall, while the integration of MANETs with satellite and terrestrial networks poses certain challenges, it also opens new avenues for improving the robustness and reach of communication systems in disaster recovery scenarios. Future research should focus on developing effective strategies for this integration, ensuring seamless interoperability while addressing potential technical challenges.

5.2.3.1 Satellite Networks Integration

Integrating MANETs with satellite networks can significantly extend the communication range in disaster-stricken areas, especially where terrestrial networks are non-functional or destroyed.

Satellite networks could provide a reliable backhaul connection for MANET nodes, enabling communication with rescue centers located far away or even in another country.

The integration, however, brings about several technical challenges. For instance, satellite communication often experiences high latency due to the large distance signals must travel. This latency can impact the performance of MANET routing protocols and hamper real-time communication. Furthermore, satellite connections often have limited bandwidth, which can be quickly exhausted when serving numerous MANET nodes.

Solutions to these challenges might involve the use of advanced routing protocols that can tolerate high latency and the implementation of bandwidth-efficient communication strategies. The development of hybrid networks that utilize both satellite and terrestrial connections could also be a viable strategy to optimize network performance.

5.2.3.2 Terrestrial Networks Integration

Incorporating terrestrial networks with MANETs can provide additional connectivity options and enhance network coverage in disaster-stricken areas. This integration can be particularly beneficial in urban settings where terrestrial network infrastructure, such as cellular towers, is available but may be overloaded due to increased demand during a disaster. However, this integration also comes with a set of challenges. For instance, the heterogeneity in the network standards and protocols between MANETs and terrestrial networks can hinder seamless communication. Additionally, terrestrial networks can also be vulnerable to the same disasters affecting the area, limiting their availability and reliability. Solutions to these challenges can include the development of adaptive protocols that can seamlessly handle diverse network standards and the use of robust, disaster-resistant infrastructure for terrestrial networks.

In conclusion, the integration of MANETs with satellite and terrestrial networks holds great promise for enhancing disaster communication coverage. However, it also presents a set of technical challenges that need to be addressed. Through ongoing research and innovation, we can develop effective strategies and solutions that leverage the strengths of each network type to achieve seamless interoperability and enhanced communication in disaster scenarios.

5.3 FUTURE DIRECTIONS IN ADVANCED ROUTING PROTOCOLS

The field of advanced routing protocols for MANETs is likely to continue evolving rapidly. This is driven by the growing demands of MANET applications, as well as the continuous advancement of network technologies. One key area of focus is likely to be the optimization of routing protocols for specific MANET applications. For instance, protocols designed for disaster recovery scenarios may prioritize robustness and resilience, while those for IoT applications may focus on energy efficiency and scalability. Moreover, as network technologies continue to evolve, new types of routing protocols may emerge. These could leverage advanced technologies like quantum computing or blockchain to enhance network security, reliability, and performance.

Furthermore, the integration of advanced routing protocols with emerging technologies like IoT and ML will continue to be a significant area of focus. This integration can unlock new capabilities for these protocols, enhancing their performance and utility in diverse MANET applications.

Emerging technologies, such as blockchain, IoT, and ML, are expected to play a significant role in shaping this evolution. The integration of these technologies with advanced routing protocols can potentially enable more secure, efficient, and versatile MANETs, significantly enhancing their utility in various applications including disaster recovery scenarios. The continuous innovation and integration of technologies, such as IoT, ML and quantum computing, are expected to drive this growth, pushing the boundaries of what is possible in network performance, security, and versatility. These advancements will undoubtedly enhance the utility of MANETs in various applications, including disaster recovery.

In disaster recovery scenarios, the need for reliable and efficient communication is paramount. The integration of these advanced technologies with MANETs can enable faster and more accurate sharing of critical information, facilitating more efficient response efforts and potentially saving lives.

Furthermore, these technologies can also assist in overcoming some of the inherent challenges in MANETs. For instance, ML algorithms can be utilized to optimize network performance under different conditions, while IoT devices can provide valuable data to aid in decision-making processes. Quantum computing, although still in its early stages, holds the potential to revolutionize aspects of data security and encryption, which are crucial in maintaining the integrity and reliability of communication networks. However, the integration of these technologies also presents new challenges and areas for further research. These include ensuring the efficient management and security of the vast amounts of data generated by IoT devices, developing robust ML algorithms that can adapt to dynamic network conditions, and understanding and harnessing the potential of quantum computing within the context of MANETs.

In conclusion, the future of MANETs in disaster recovery looks promising, with ongoing advancements in technology expected to continually improve their performance, versatility, and reliability. By continuing to explore and innovate in this field, we can look forward to more robust and effective communication networks that can significantly enhance disaster recovery efforts.

5.4 FUTURE DIRECTIONS IN MANET FOR DISASTER RECOVERY

Numerous research areas in MANETs for disaster recovery emerge as potential directions for future exploration and development. These include the ongoing development of energy-efficient networking strategies, robust hardware for harsh environments, secure routing protocols, and solutions for network setup complexity. Moreover, enhancing interoperability between MANETs and other network types remains a significant challenge that warrants further investigation. the deployment of MANETs in simulated and real-world disaster scenarios can provide valuable insights into their operational strengths and weaknesses. These experiences can inform future development efforts, leading to more robust and versatile networks for disaster recovery.

5.4.1 Network Security

Despite considerable advancements, network security remains a significant challenge in MANETs. The dynamic and decentralized nature of these networks makes them vulnerable to various security threats, including node capture, denial-of-service attacks, and data tampering. Future research efforts in this area could focus on developing more robust security protocols and mechanisms, possibly leveraging emerging technologies, such as blockchain and ML.

5.4.2 Quality of Service

Ensuring a high QoS is crucial in disaster recovery scenarios, where the timely and reliable transmission of data can have life-saving implications. Future research could explore ways to enhance the QoS in MANETs, such as through the development of more efficient routing protocols, the integration with high-speed wireless technologies like 5G, or the application of ML techniques for network optimization.

5.4.3 Scalability

As the potential applications of MANETs continue to grow, so does the need for these networks to scale efficiently. Future research in this area could focus on developing strategies and protocols that allow MANETs to handle a larger number of nodes without a significant drop in performance. This could involve exploring new network architectures or integrating MANETs with other networking technologies such as the IoT.

5.4.4 Energy Efficiency

Given the often-limited power resources in disaster recovery scenarios, improving the energy efficiency of MANETs is a critical research direction. This could involve the development of energy-aware routing protocols, the use of energy-efficient network hardware, or the application of power management techniques such as sleep scheduling.

In conclusion, while significant progress has been made in the field of MANETs for disaster recovery, numerous opportunities for further research and development remain. By continuing to explore these opportunities, we can strive to make these networks even more robust, efficient, and versatile, thereby enhancing their capacity to support disaster recovery efforts.

5.5 PRACTICAL DEPLOYMENT CHALLENGES AND SOLUTIONS

Practical deployment of MANETs in disaster recovery scenarios can face various challenges. These might include network setup complexity, energy constraints, unpredictable environmental factors, and security concerns. The complexity of setting up a MANET in a disaster scenario can be daunting, particularly given the urgency and stress associated with such situations. Simplifying the deployment process through automated setup procedures or user-friendly interfaces can significantly alleviate this challenge.

Energy constraints are another key challenge. Nodes in a MANET often rely on limited power sources, and in a disaster situation, recharging or replacing these power sources may not be feasible. Implementing energy-efficient networking strategies, such as power-saving routing protocols or sleep scheduling, can help to extend the operational lifespan of the network. Unpredictable environmental factors, such as harsh weather conditions or physical obstacles, can also hinder the deployment and operation of MANETs. To address this, robust hardware that can withstand harsh conditions and sophisticated routing protocols that can adapt to changes in the network environment may be necessary.

On the other hand, security is a crucial concern in the deployment of MANETs. Ensuring data integrity and confidentiality is paramount, particularly when sensitive information is being transmitted. Incorporating secure routing protocols, data encryption techniques, and node authentication strategies can help to enhance the security of the network.

Addressing these challenges requires a comprehensive approach that combines technological innovation, user-centric design, and rigorous testing. By doing so, we can enhance the practicality and effectiveness of MANETs in supporting disaster recovery efforts. While the challenges faced during the practical deployment of MANETs in disaster recovery scenarios are significant, ongoing research and technological advancements continue to provide promising solutions. The development of user-friendly setup procedures, energy-efficient networking strategies, robust hardware, secure routing protocols, and interoperability solutions are all critical steps towards overcoming these challenges.

Future research and development efforts need to continue focusing on these areas to further enhance the robustness, efficiency, and versatility of MANETs in disaster recovery scenarios. By doing so, we can create more effective communication networks that can significantly improve disaster recovery efforts.

5.6 CASE STUDIES AND REAL-WORLD APPLICATIONS

In the real-world applications section, we can discuss specific instances where MANET technologies have been deployed for disaster recovery. This may include various natural disasters like earthquakes, floods, and fires where traditional communication infrastructure may have been damaged or destroyed. The section can highlight how MANETs provided critical communication channels in such situations, allowing for effective coordination of rescue efforts, rapid dissemination of information, and support of medical and other emergency services. Case studies from different geographic locations and disaster types can be included to demonstrate the versatility and reliability of MANETs in diverse scenarios. The real-world applications can also serve as practical examples to illustrate the theoretical concepts and technical aspects discussed in previous sections.

5.6.1 Earthquake Response

One example of a real-world application of MANETs can be found in the aftermath of severe earthquakes. Traditional communication infrastructures are often severely damaged during such events, crippling the flow of vital information. In such scenarios, MANETs have been deployed to establish ad hoc communication networks that enable rescue teams to coordinate their efforts more effectively. The flexible and self-organizing nature of MANETs allows them to provide reliable communication channels even in the face of such severe disruptions.

5.6.2 Flood Relief

Floods often lead to widespread damage to communication infrastructures, impeding efficiency of rescue and recovery operations. In such cases, MANETs have proven to be invaluable for establishing temporary communication networks. These networks facilitate timely dissemination of critical information, aiding in evacuation efforts, coordinating relief operations, and assisting in locating and rescuing stranded individuals.

5.6.3 Wildfire Management

In the context of wildfires, rapid and reliable communication is crucial to coordinate firefighting efforts, evacuate at-risk populations, and monitor the progression of the fires. MANETs have been used in these scenarios to provide robust and adaptable communication networks that support these operations. By enabling real-time data sharing among firefighting teams and support services, MANETs have played a crucial role in enhancing the effectiveness of wildfire management strategies.

5.6.4 Other Disaster Scenarios

MANETs have also been used in various other disaster scenarios, including industrial accidents, nuclear incidents, and in the aftermath of terrorist attacks. In each case, the ability of MANETs to quickly establish a communication network in the absence of traditional infrastructure has proven crucial in managing the disaster and mitigating its impact.

In conclusion, the case studies and real-world applications of MANETs in disaster recovery highlight the significant role these networks can play in supporting emergency response efforts. They provide practical evidence of the potential of MANETs to enhance communication, coordination, and decision-making processes in critical situations, thereby contributing to more effective disaster management strategies.

5.7 SUCCESS STORIES OF MANETS IN DISASTER RECOVERY

Present case studies where MANETs have been successfully deployed in disaster recovery scenarios, highlighting the impact on emergency response efforts and lessons learned. MANETs have proven to be invaluable tools in disaster recovery scenarios. Several case studies showcase their successful deployment and the positive impact on emergency response efforts. In the aftermath of the devastating earthquake in Haiti in 2010, MANETs were deployed to establish communication networks in areas where traditional infrastructure was destroyed. The use of MANETs facilitated timely and effective communication among rescue teams, greatly enhancing their coordination efforts. Similarly, during the 2011 tsunami in Japan, MANETs were used to create temporary communication networks that enabled rescue teams to coordinate their efforts more effectively. Despite the widespread destruction of communication infrastructure, the use of MANETs ensured that vital information could be exchanged quickly and efficiently.

In both cases, the deployment of MANETs in disaster scenarios underscored their potential in enhancing communication and coordination among rescue teams. The lessons learned from these experiences include the need for robust, easy-to-deploy MANET technologies that can withstand challenging conditions and the importance of training emergency response teams in the use of such technologies.

These case studies demonstrate the significant impact of MANETs on disaster recovery efforts. Moving forward, it is essential to continue research and development in this area, improving the robustness, efficiency, and ease-of-use of MANETs to further enhance their effectiveness in disaster recovery scenarios.

Another successful deployment of MANETs was during the 2013 Colorado floods. With traditional communication infrastructure severely damaged, rescue and emergency services were able to utilize MANETs to establish a communication network [89]. This allowed for the effective coordination of rescue efforts, efficient dissemination of crucial information, and real-time data sharing which significantly enhanced the effectiveness of rescue operations.

During the 2018 California wildfires, MANETs were employed for real-time data sharing among firefighting teams and support services [89]. This greatly improved the coordination of firefighting efforts and the evacuation of at-risk populations, demonstrating the critical role of MANETs in managing such disasters [90].

MANETs have also been used in industrial accidents and nuclear incidents. For instance, following the Chernobyl nuclear disaster, MANETs were used to establish communication networks that enabled efficient coordination among response teams. This helped manage the situation more effectively, limiting further damage and exposure to radiation. These case studies underscore the versatility and reliability of MANETs in diverse disaster scenarios. They highlight the critical role that MANETs can play in enhancing communication, coordination, and decision-making processes in critical situations, contributing to more effective disaster management strategies. The lessons learned from these experiences emphasize the importance of further research and development in this field. As we continue to innovate and improve MANET technology, we can expect to see even more robust and effective communication networks that can significantly enhance disaster recovery efforts [91].

5.8 SUMMARY

This chapter provided an advanced concepts and future directions in MANETs for disaster recovery. It explored the historical progression of MANETs and their application in disaster recovery scenarios. The study also discussed potential future trends, including the integration with other emerging technologies such as 5G, Wi-Fi 6, and the IoT. Energy-efficient networking within MANETs, the development of advanced routing protocols, and the use of ML for dynamic route optimization are highlighted as key areas of focus. However, the chapter also notes the challenges associated with these advancements, including security concerns and the need for new protocols and standards.

5.9 KEY TERMS

MANET, Disaster Recovery, Network Design, Evolution of MANET Technologies, Emerging Wireless Technologies, Energy-Efficient Networking, Advanced Routing Protocols, ML-Based Routing.

5.10 REVIEW QUESTIONS

1. What are the key historical developments and technological advancements of MANETs?

2. How do emerging wireless technologies like 5G and Wi-Fi 6 potentially enhance the capabilities of MANETs in disaster recovery scenarios?

3. What is the importance of energy-efficient networking in the context of MANETs?

4. How have advanced routing protocols evolved to manage the dynamic nature of MANETs and what challenges do they face?

5. How can integrating ML with advanced routing protocols potentially enhance the performance of MANETs and what challenges might arise with such integration?

6. What are some potential applications of MANETs in disaster recovery scenarios and what are the associated benefits and challenges?

7. How can emerging technologies like IoT and quantum computing contribute to the advancement of MANETs in disaster recovery scenarios?

8. How can the integration of blockchain technology enhance the security measures within MANETs, and what potential challenges could this integration present?

9. Considering the advancements in quantum computing, how might this technology revolutionize data encryption and security within MANETs, and what potential obstacles might be encountered in this integration?

5.11 MINI PROJECTS

Project: Enhancing MANET Performance in Disaster Recovery Scenarios

Objective: The goal of this mini project is to develop a sound knowledge and understanding of MANET by developing a theoretical model for MANET-based disaster recovery scenarios.

Activities:

1. Research and Analysis: Researching on ANET applications in disaster recovery scenarios. Develop a good understanding of key principles, technologies, and protocols associated with

MANETs. Analyze the challenges of deploying MANETs in disaster-stricken areas.

2. Model Development: Develop a theoretical model for a MANET that can be deployed in a disaster-stricken area. The model should consider the factors like dynamic topology, limited power resources, and the need for rapid deployment.

3. Integration with Emerging Technologies: Explore the integration of emerging technologies like IoT, 5G, and ML to enhance the performance of MANET that you developed in item 2 above.

4. Energy-Efficient Networking: Propose a technique to make your MANET model energy-efficient. Consider strategies such as implementing power-efficient routing protocols and using energy-efficient hardware.

5. Advanced Routing Protocol: Study some advanced routing protocols such as DSR and AODV. Propose a method to enhance the performance of the proposed MANET model.

6. Research Report: Write a report summarizing your research findings. Present the report to your peers and instructors.

Expected Outcome: Students will gain a deep understanding of MANETs and system performance evaluation in disaster recovery scenarios. They also explore the integration of MANETs with emerging technologies and energy-efficient networking.

CHAPTER 6

Conclusion and Discussion

LEARNING OBJECTIVES

After reading and completing this chapter, you will be able to:

- Discuss the recent developments in MANETs.

- Discuss the deployment scenarios of MANETs in disaster recovery.

- Identify and discuss three key future research directions in MANET disaster recovery.

6.1 DEPLOYMENT OF MANETS IN DISASTER RECOVERY

Deploying wireless networks for Mobile Ad Hoc Networks (MANETs) in disaster recovery scenarios presents a formidable challenge due to the necessity for high data rates to facilitate effective communication. Establishing a reliable communication infrastructure is of paramount importance for communicating with individuals in disaster recovery zones. Drawing from prior

DOI: 10.1201/9781032700571-6 **101**

research, this book has opted for the utilization of the 802.11b, a second-generation wireless network protocol, within its simulation. The book has undertaken a comprehensive examination of the impact of individuals' varying movement speeds within a disaster-stricken area. The analytical findings underscore the substantial influence of mobility, particularly when node density is elevated, on the throughput of Wi-Fi links.

6.2 DENSE NETWORK SCENARIO

When disasters strike, disaster relief communications become a multifaceted challenge. Communication infrastructures, including cellular towers, may be demolished, further exacerbated when high-density victim populations are involved. This book delves into an exploration of the impact of nodes, symbolizing disaster victims in recovery areas. The findings reveal that as the number of victims in an area increase, communication throughput diminishes. Nevertheless, there are scenarios where certain nodes boast a higher number of neighbors, while others have fewer due to limited node mobility throughout the area. The graphical representations of these effects vary, with throughput exhibiting fluctuations, particularly for nodes emulating the velocities of individuals walking and those in vehicles, as node density increases. Notably, nodes with high velocities have a profound influence on average delay results, especially under conditions of maximum node density.

6.3 LARGE NETWORK SCENARIO

This book has primarily concentrated on assessing the impact of GWRSs on MANET performance, employing the scenario of a natural disaster occurring in Loja City, Ecuador. Within a disaster-stricken area, nodes are expected to exhibit diverse velocities. This book has considered various node mobility parameters to replicate the movement patterns of individuals during disaster recovery. The endeavor of deploying communication networks for disaster recovery poses a formidable challenge for network engineers. They must meticulously determine the optimal number of nodes, the appropriate velocities for nodes within the disaster area, and strategically position gateways to attain the desired performance and coverage. The results elucidated in Chapter 5 underscore the importance of maintaining a balanced ratio between the number of gateways and the number of nodes to mitigate issues like packet delay and packet loss. This equilibrium is pivotal, particularly given the pivotal role of communication in the aftermath of a disaster. The analyses conducted in this book have taken into consideration the real-world ratio of individuals involved in the Loja City disaster scenario.

6.4 HIGH NODE MOBILITY

This book conducted an in-depth examination of the range of velocities exhibited by individuals during disaster scenarios. This comprehensive investigation encompassed the study of various scenarios, including (1) stationary individuals, (2) pedestrians moving at both minimum and maximum velocities, (3) individuals in vehicles operating at both minimum and maximum speeds, and (4) individuals momentarily halting at specific locations for varying durations.

Furthermore, this book has made a significant contribution by scrutinizing the impact of node mobility on MANET performance within disaster recovery contexts. The research delved into multifaceted challenges inherent to disaster recovery, including (a) obstacles, (b) nodes entering or exiting the network unexpectedly, and (c) signal propagation dynamics. Drawing from the simulation results it is evident that node mobility plays a pivotal role in shaping network performance. This book stands as a valuable reference tool, offering guidance to network engineers involved in the planning and deployment of communication networks in disaster-stricken areas. When embarking on the deployment of communication networks in disaster scenarios, network engineers must make strategic decisions regarding the installation of optimal networking equipment capable of sustaining high-performance operation.

6.5 CONCLUSION

This chapter has provided comprehensive deployment guidelines for utilizing MANETs in disaster recovery areas, drawing from the in-depth analysis of the results presented in Chapter 4. Additionally, it has proposed potential avenues for extending this research. The findings underscore the pivotal roles played by gateway selection schemes, routing protocols, and node mobility in enhancing network performance within disaster recovery contexts. This study has shed light on the critical network attributes that must be considered when designing solutions for disaster recovery. The analytical approach in this research incorporated elements such as high mobility, variable node density, and diverse pause times. In summary, this research makes a valuable contribution by offering insights into MANET performance analysis within the context of network disaster recovery, with a specific focus on GWRSs.

As for future directions, several recommendations are put forth. These include evaluating performance using mobility models that closely emulate real-world disaster recovery scenarios, exploring dedicated simulation tools that align with the selected mobility models, incorporating network lifetime as a performance metric, given the significance of battery life in disaster recovery situations, and investigating the integration of Voronoi diagrams into GWRS schemes in conjunction with various mobility models. The forthcoming section outlines a summary of the advancements envisioned for this research.

6.6 FUTURE RESEARCH DIRECTIONS

This book delves into the realm of MANET performance within disaster recovery areas, prompted by the transformative influence of the Internet on modern life. In an era where wireless connectivity has become integral to daily existence, disruptions in network communication due to natural disasters pose substantial challenges. The increasing frequency of such disasters, coupled with the vulnerability of telecommunication infrastructures, underscores the critical need to enhance network communication during disaster recovery efforts.

MANETs, characterized by their ability to be rapidly deployed without reliance on existing network systems, emerge as a potent solution in disaster-stricken regions. Comprising mobile nodes with varying velocities within the affected area, MANETs facilitate packet forwarding from source to destination through a relay mechanism involving neighboring nodes. Additionally, certain nodes with external network connectivity serve as gateways, enabling MANET nodes to communicate with external networks, including the Internet.

The simulation research methodology is employed in evaluating the performance of MANET in disaster recovery scenarios. This approach, found to be apt for this study, allows for the generalization of measurement results and the estimation of performance metrics. This book introduces three distinctive contributions, which have been elaborated upon in preceding chapters:

- **Enhancement of gateway selection schemes:** A central focus of this research is the development of gateway selection schemes that optimize traffic load balancing among gateways. Simulations, conducted under both generic and disaster-specific conditions resembling Loja City, Ecuador, reveal the necessity for efficient gateway selection schemes to manage high-density traffic loads effectively. The OMNET++ simulation tool, coupled with the INETMANET framework, was instrumental in evaluating scheme performance based on metrics such as mean throughput, packet end-to-end delay, packet drop ratio, and packet sent rate. Notably, results indicated fluctuations in scenarios featuring limited gateways and high node density.

- **Improving MANET performance using routing selection schemes:** The book emphasizes the importance of routing selection schemes tailored to disaster recovery scenarios. Traditional routing protocols like AODV and DSDV employ broadcasting algorithms for route discovery, leading to network flooding and intricate route selection processes. In disaster recovery situations, simultaneous Internet connection attempts by numerous victims can overwhelm the network. The proposed routing selection scheme significantly enhances MANET performance compared to conventional AODV and DSDV schemes. Even with high node density and dynamic topology changes due to node mobility, the proposed scheme exhibits superior performance in terms of packet throughput, packet end-to-end delay, packet drop rate, and packet send rate.

- **Node mobility effects in disaster recovery areas:** An exploration of node mobility's impact in disaster recovery zones is another key aspect of this research. Through simulations, node behaviors are scrutinized, revealing that mobile devices can exhibit high mobility during disaster responses and evacuation efforts. Results indicate that node speed introduces complexities in maintaining communication, as neighboring nodes can suddenly appear or relocate randomly. The research employs the GWRS (Gateway-Based Routing Selection) scheme across various node speeds, distinguishing between static individuals, pedestrians, and individuals in vehicles during disaster recovery. Findings demonstrate that GWRS delivers high throughput, even in scenarios with high node density and variable node speeds. However, when individuals in vehicles travel at maximum velocity, although average packet delay increases, throughput remains substantial, exceeding 0.7 Mbps. This underscores GWRS's capacity to provide ample bandwidth in disaster recovery contexts. Overall, the heterogeneous velocity of nodes within disaster areas exerts a discernible influence on MANET performance.

In conclusion, this research serves as a valuable reference for understanding and optimizing MANET performance in disaster recovery settings. The outlined recommendations for future developments include evaluating performance using more realistic mobility models, exploring dedicated simulation tools, considering network lifetime as a crucial metric, and integrating Voronoi diagrams into GWRS schemes, thereby providing a roadmap for further advancements in this field.

6.7 KEY TERMS

MANET, Dense Network, Velocity, Large network, High Mobility, Throughput, Delay, Gateway Selection Schemes, Routing Selection Schemes, Node Mobility, Load Balancing, Performance Metrics, Simulation Model, GWRS, Network Lifetime, Voronoi Diagrams, Disaster Recovery, Packet Forwarding.

6.8 REVIEW QUESTIONS

1. List and discuss two key challenges of deploying MANETs in disaster recovery scenarios.

2. Explain how gateway selection schemes can be used to improve load balancing in MANETs?

3. Explain how node mobility affects the performance of MANETs in disaster recovery areas?

4. Explain why computer simulation is suitable for evaluating the performance of MANETs in disaster recovery?

5. Identify and discuss three key performance metrics commonly used is system evaluation.

6. What deployment guidelines should network engineers follow when setting up MANETs in disaster recovery areas? Explain why is maintaining a balanced gateway-to-node ratio crucial?

6.9 MINI PROJECTS

The following mini projects aim to provide a deeper understanding of the topics covered in this chapter using a literature review and empirical study.

1. Conduct an in-depth review of the literature focusing on deploying MANETs in disaster recovery scenarios. Read 15 to 20 recent relevant journal/conference papers to identify the key researchers and their main contributions on MANET deployment in disaster scenarios. You can use Table 5.1 as a template to record your findings.

2. Conduct an in-depth review of the literature on various MANET gateway selection schemes suitable for disaster recovery. Read about 20 recent relevant journal/conference papers to identify the key researchers and their main contributions on above gateway selection areas. You can use Table 6.1 as a template to record your findings.

3. Develop a simulation model (using OMNET++ or any other credible simulation tools) to study the impact of node mobility on a typical MANET system performance. Measure key performance metrics such as throughput, delay, and load balancing.

4. Develop a simulation model to study the performance of a MANET routing selection scheme tailored for disaster recovery. Measure the key performance metrics (throughput, delay, and load balancing) and analyze the system performance to make a meaningful conclusion.

5. Develop a simulation model to study the planning of the deployment of MANETs in a disaster-stricken area. Simulate the network to determine the optimal node placement and gateway positioning to achieve the desired coverage and performance.

TABLE 6.1   Leading Researchers and Their Contributions to MANET
Deployments

Researcher	Contribution	Year	Description/Key Concept

6. Develop a MANET simulation model to study the impact of various network parameters on system performance (throughput, end-to-end delay, and packet drop ratio).

7. Develop a short research proposal (2 to 4 pages) on MANETs for disaster recovery. Define research objectives, methods, and expected outcomes, considering aspects like realistic mobility models and advanced routing schemes.

References

[1] B. Nancharaiah, and B. Chandra Mohan, "The performance of a hybrid routing intelligent algorithm in a mobile ad hoc network," *Comput. Electr. Eng.,* vol. 40, no. 4, pp. 1255–1264, May 2014.

[2] J. Ebenezer, "A mobility model for MANET in large scale disaster scenarios," in *A Mobility Model for MANET in Large Scale Disaster Scenarios,* IEEE, New York, 2014, pp. 59–64.

[3] D. G. Reina, M. Askalani, S. L. Toral, F. Barrero, E. Asimakopoulou, and N. Bessis, "A survey on multihop ad hoc networks for disaster response scenarios," *Int. J. Distrib. Sens. Netw.,* vol. 11, no. 10, p. 647037, Oct. 2015.

[4] N. Mahiddin, N. Sarkar, and B. Cusack, "An internet access solution: MANET routing and a gateway selection approach for disaster scenarios," *Rev. Socionetw.,* vol. 11, no. 1, pp. 47–64, Jun. 2017.

[5] D. G. Reina, J. M. León-Coca, S. Toral, E. Asimakopoulou, F. Barrero, P. Norrington, and N. Bessis, "Multi-objective performance optimization of a probabilistic similarity/dissimilarity-based broadcasting scheme for mobile ad hoc networks in disaster response scenarios," *Soft Comput.,* vol. 18, no. 9, pp. 1745–1756, Sept. 2014.

[6] D. Helen, and D. Arivazhagan, "Applications, advantages and challenges of ad hoc networks," *J. Acad. Ind. Res.,* vol. 2, no. 8, pp. 453–457, Jan. 2014.

[7] L. Raja, and S. S. Baboo, "An overview of MANET: Applications, attacks and challenges abstract," *Int. J. Comput. Sci. Mob. Comput.,* vol. 3, no. 1, pp. 408–417, Jan. 2014.

[8] K. Tweed, "How solar is playing a role in Nepal's disaster relief," *greentechmedia.com,* 2015. [Online]. Available: https://www.greentechmedia.com/articles/read/role-for-solar-in-nepal-disaster-relief#gs. JVJFYnM.

[9] T. Love, "Solar power is being used as disaster relief. Here's how," World. Econ. Forum., May 2018. [Online]. Available: https://www. weforum.org/stories/2018/05/how-solar-power-is-impacting-natural-disaster-relief/.

[10] Y. Miao, Z. Sun, F. Yao, N. Wang, and H. S. Cruickshank, "Study on research challenges and optimization for internetworking of Hybrid MANET," in *Personal Satellite Services*, Springer, New York, 2013, pp. 90–101.

[11] R.-U. Zaman, K.-U.-R. Khan, and A. V. Reddy, "A review of gateway load balancing strategies in Integrated Internet-MANET," in *Proceedings of 2009 IEEE International Conference on Internet Multimedia Services Architecture and Applications (IMSAA'09)*, IEEE, Bangalore, 2009, pp. 1–6.

[12] S. H. Bouk, I. Sasase, S. H. Ahmed, and N. Javaid, "Gateway discovery algorithm based on multiple QoS path parameters between mobile node and gateway node," *J. Commun. Netw.*, vol. 14, no. 4, pp. 434–442, Aug. 2012.

[13] R. Manoharan, and S. Mohanalakshmie, "A Trust Based Gateway Selection Scheme for Integration of MANET with Internet," in *Proceedings of IEEE-International Conference on Recent Trends in Information Technology (ICRTIT'11)*, IEEE, Chennai, 2011, pp. 543–548.

[14] S. Prabhavat, H. Nishiyama, and N. Ansari, "On load distribution over multipath networks," *IEEE Commun. Surv. Tutor.*, vol. 14, no. 3, pp. 662–680, Sept. 2012.

[15] A. Oliveira, P. Z. Sun, M. Monier, and P. Boutry, "On optimizing hybrid ad-hoc and satellite networks—the MONET approach," in *Proceedings of Future Network and Mobile Summit*, University of Surrey, Guildford, 2010, pp. 1–8.

[16] J. H. Zhao, X. Z. Yang, and H. W. Liu, "Load-balancing strategy of multi-gateway for ad hoc Internet connectivity," in *Proceedings of International Conference on Information Technology: Coding and Computing (ITCC'05)*, IEEE, Las Vegas, NV, 2005, vol. 2, pp. 592–596.

[17] C.-K. Toh, "Associativity-based routing for ad-hoc mobile networks," *Kluwer J. Wirel. Pers. Commun.*, vol. 4, pp. 103–139, Nov. 1997.

[18] V. D. Park, and M. S. Corson, "A highly adaptive distributed routing algorithm for mobile wireless networks," in *Proceedings of IEEE Conference on Computer Communications (INFOCOM'97)*, IEEE, Kobe, 1997, vol. 3, pp. 1405–1413.

[19] S. Murthy, and J. J. Garcia-Luna-Aceves, "An efficient routing protocol for wireless networks," *Mob. Networks Appl.*, vol. 1, no. 3, pp. 183–197, Oct. 1996.

[20] C.-C. Chiang, H.-K. Wu, W. Liu, and M. Gerla, "Routing in clustered multihop, mobile wireless networks with fading channel," in *Proceedings of IEEE SICON*, 1997, vol. 97, pp. 197–211.

[21] J. N Al-karaki, and A. E. Kamal, "Efficient virtual-backbone routing in mobile ad hoc networks," *Comput. Netw.*, Vol. 52, no. 2, pp. 327–350, Feb. 2008.

[22] P. Guangyu, M. Geria, and X. H. X. Hong, "LANMAR: Landmark routing for large scale wireless ad hoc networks with group mobility," in *Proceedings of 2000 First Annual Workshop On Mobile Ad Hoc Networking and Computing Conference, MobiHOC (Cat. No.00EX444)*, IEEE, Boston, MA, 2000, pp. 11–18.

[23] G. Pei, M. Gerla, and T.-W. Chen, "Fisheye state routing: A routing scheme for ad hoc wireless networks," in *Proceedings of IEEE International Conference on Pervasive Computing and Communications*, IEEE, Toronto, ON, 2000, vol. 1, pp. 70–74.

[24] C.-C. Yang, and L.-P. Tseng, "Fisheye zone routing protocol: A multi-level zone routing protocol for mobile ad hoc networks," *Comput. Commun.*, vol. 30, no. 2, pp. 261–268, Jan. 2007.

[25] G. Mesut, U. Sorges, and I. Bouazizi, "ARA—The ant-colony based routing algorithm for MANETs," in *Proceeding of the International Conference on Parallel Processing Workshops (ICPPW'02)*, IEEE Computer Society, Washington, DC, 2002, p. 79.

[26] L. Rosati, M. Berioli, and G. Reali, "On ant routing algorithms in ad hoc networks with critical connectivity," *Ad Hoc Netw.*, vol. 6, no. 6, pp. 827–859, Aug. 2008.

[27] S. Rajagopalan, and C. C. Shen, "ANSI: A swarm intelligence-based unicast routing protocol for hybrid ad hoc networks," *J. Syst. Archit.*, vol. 52, no. 8–9, pp. 485–504, Aug. 2006.

[28] J. Wang, E. Osagie, P. Thulasiraman, and R. K. Thulasiram, "HOPNET: A hybrid ant colony optimization routing algorithm for mobile ad hoc network," *Ad Hoc Netw.*, vol. 7, no. 4, pp. 690–705, Jun. 2009.

[29] L. R. Reddy, and S. V Raghavan, "SMORT: Scalable multipath on-demand routing for mobile ad hoc networks," *Ad Hoc Netw.*, vol. 5, no. 2, pp. 162–188, Mar. 2007.

[30] T.-W. Chen, and M. Gerla, "Global state routing: A new routing scheme for ad-hoc wireless networks," in *Proceedings of IEEE International Conference on Communications Conference Record Affiliated with SUPERCOMM'98*, IEEE, Atlanta, GA, 1998, vol. 1, pp. 171–175.

[31] C. E. Perkins, and E. M. Royer, "Ad-hoc on-demand distance vector routing," in *Proceedings WMCSA'99. Second IEEE Workshop on Mobile Computing Systems and Applications*, IEEE, New Orleans, LA, 1999, pp. 99–100.

[32] J. Raju, and J. Garcia-Luna-Aceves, "A new approach to on-demand loop-free multipath routing," in *Proceedings Eight International Conference on Computer. Communications and Networks (Cat. No.99EX370)*, IEEE, Boston, MA,1999, pp. 522–527.

[33] J. Garcia-Luna-Aceves, and M. Spohn, "Source-tree routing in wireless networks," in *Proceedings of the Seventh Annual International Conference on Network Protocols (ICNP '99)*, IEEE Computer Society, Toronto,1999, p. 273.

[34] S. Radhakrishnan, G. Racherla, C. N. Sekharan, N. S. V Rao, and S. G. Batsell, "DST-A routing protocol for ad hoc networks using distributed spanning trees," in *Proceedings of IEEE Wireless Communications and Networking Conference, WCNC'99*, IEEE, New Orleans, LA, 1999, vol. 3, pp. 1543–1547.

[35] M. Gerla, "IPv6 flow handoff in ad hoc wireless networks using mobility prediction," in *Proceedings of* Seamless Interconnection for Universal Services. Global Telecommunication Conference (GLOBECOM'99), (Cat. No.99CH37042), 1999, IEEE, Rio de Janeiro, vol. 1a, pp. 271–275.

[36] G. Aggelou, and R. Tafazolli, "RDMAR: A bandwidth-efficient routing protocol for mobile ad hoc networks," in *Proceedings of the 2nd ACM International Workshop on Wireless Mobile Multimedia*, Seattle, Washington, DC, 1999, pp. 26–33.

[37] Q. Xue, and A. Ganz, "Ad hoc QoS on-demand routing (AQOR) in mobile ad hoc networks," *J. Parallel Distrib. Comput.*, vol. 63, no. 2, pp. 154–165, Feb. 2003.

[38] C. Sengul, and R. Kravets, "Bypass routing: An on-demand local recovery protocol for ad hoc networks," *Ad Hoc Netw.*, vol. 4, no. 3, pp. 380–397, May 2006.

[39] N. Nikaein, H. Labiod, and C. Bonnet, "DDR-Distributed dynamic routing algorithm for mobile ad hoc networks," in *Proceedings of First Annual Workshop on Mobile and Ad Hoc Networking and Computing. MobiHOC*, Boston, MA, 2000, pp. 19–27.

[40] C. W. Ahn, "Gathering-based routing protocol in mobile ad hoc networks," *Comput. Commun.*, vol. 30, no. 1, pp. 202–206, Dec. 2006.

[41] D. B. Johnson, and D. a Maltz, "DSR: The dynamic source routing protocol for multi-hop wireless ad hoc networks," in *Computer Science, Department at Carnegie Mellon University,* Addison-Wesley, Boston, MA, 2001, vol. 5, pp. 139–172.

[42] Y.-H. Wang, and C.-F. Chao, "Dynamic backup routes routing protocol for mobile ad hoc networks," *Inf. Sci. (Ny).,* vol. 176, no. 2, pp. 161–185, Jan. 2004.

[43] J. Boice, J. J. Garcia-Luna-Aceves, and K. Obraczka, "Combining on-demand and opportunistic routing for intermittently connected networks," *Ad Hoc Netw.,* vol. 7, no. 1, pp. 201–218, Jan. 2009.

[44] G. Wang, D. Turgut, L. Bölöni, Y. Ji, and D. C. Marinescu, "Improving routing performance through m-limited forwarding in power-constrained wireless ad hoc networks," *J. Parallel Distrib.* Comput., vol. 68, no. 4, pp. 501–514, Apr. 2008.

[45] G. Wang, Y. Ji, and D. C. Marinescu, "A routing protocol for power constrained networks with asymmetric links," in *Proceedings of the 1st ACM International Workshop on Performance Evaluation of Wireless ad hoc, Sensor, and Ubiquitous Networks (PE-WASUN '04),* Venezia, 2004, pp. 69–76.

[46] N.-C. Wang, Y.-F. Huang, and J.-C. Chen, "A stable weight-based on-demand routing protocol for mobile ad hoc networks," *Inf. Sci. (Ny).,* vol. 177, no. 24, pp. 5522–5537, Dec.2007.

[47] M. Yu, A. Malvankar, W. Su, and S. Y. Foo, "A link availability-based QoS-aware routing protocol for mobile ad hoc sensor networks," *Comput. Commun.,* vol. 30, no. 18, pp. 3823–3831, Dec. 2007.

[48] X. Xie, W. Gang, W. Keping, W. Gang, and J. Shilou, "Link reliability based hybrid routing for tactical mobile ad hoc network," *J. Syst. Eng. Electron.,* vol. 19, no. 2, pp. 259–267, Jan. 2008.

[49] G. I. Ivascu, S. Pierre, and A. Quintero, "QoS routing with traffic distribution in mobile ad hoc networks," *Comput. Commun.,* vol. 32, no. 2, pp. 305–316, Feb. 2009.

[50] A. Boukerche, S. K. Das, and A. Fabbri, "Analysis of a randomized congestion control scheme with DSDV routing in ad hoc wireless networks," *J. Parallel Distrib. Comput. (Special Issue Wirel. Networks),* vol. 61, no. 7, pp. 967–995, Jul. 2001.

[51] P. Jacquet, P. Mühlethaler, T. H. Clausen, A. Laouiti, A. Qayyum, and L. Viennot, "Optimized link state routing protocol for ad hoc networks," in *Proceedings of IEEE International Multi Topic Conferences (INMTC'01),* IEEE, Lahore, 2001, pp. 62–68.

[52] L. Villasenor-Gonzalez, Y. Ge, and L. Lamont, "HOLSR: A hierarchical proactive routing mechanism for mobile ad hoc networks," *IEEE Commun. Mag.,* vol. 43, no. 7, pp. 118–125, Jul. 2005.

[53] A. Munaretto, and M. Fonseca, "Routing and quality of service support for mobile ad hoc networks," *Comput. Netw.*, vol. 51, no. 11, pp. 3142–3156, Aug. 2007.

[54] J. Eisbrener *et al.*, "Recycled path routing in mobile ad hoc networks," *Comput. Commun.*, vol. 29, no. 9, pp. 1552–1560, May 2006.

[55] R. Beraldi, L. Querzoni, and R. Baldoni, "A hint-based probabilistic protocol for unicast communications in MANETs," *Ad Hoc Netw.*, vol. 4, no. 5, pp. 547–566, Sept. 2006.

[56] R. Dube, C. D. Rais, K.-Y. Wang, and S. K. Tripathi, "Signal stability-based adaptive routing (SSA) for Ad hoc mobile networks," *IEEE Pers. Commun.*, vol. 4, no. 1, pp. 36–45, Feb. 1997.

[57] W. Su and M. Gerla, "IPv6 flow handoff in ad hoc wireless networks using mobility prediction," in *Proceedngs of IEEE Global Telecommunications Conferences (Globecom'99)*, IEEE, Rio de Janeiro, 1999, pp. 0–4.

[58] C. E. Perkins, M. Park, and E. M. Royer, "Ad-hoc on-demand distance vector routing," in Proceedings of *IEEE Workshop on Mobile Computing Systems and Applications (WMCSA'99)*, IEEE, New Orleans, LA, 1999, pp. 99–100.

[59] M. Mosko and C. E. Perkins, "A new approach to on-demand loop-free routing in networks using sequence numbers," *Comput. Netw.*, vol. 50, pp. 1599–1615, Jul. 2006.

[60] C. W. Yu, T.-K. Wu, and R. H. Cheng, "A low overhead dynamic route repairing mechanism for mobile ad hoc networks," *Comput. Commun.*, vol. 30, no. 5, pp. 1152–1163, Mar. 2007.

[61] H. Rangarajan, and J. J. Garcia-Luna-Aceves, "Efficient use of route requests for loop-free on-demand routing in ad hoc networks," *Comput. Netw.*, vol. 51, pp. 1515–1529, Apr. 2007.

[62] C. E. Perkins, and P. Bhagwat, "Highly dynamic destination -sequenced distance-vector routing (DSDV) for mobile computers," in *Proceedings of the Conference of the ACM Special Interest Group on Data Communication (ACM SIGCOMM'94)*, London, 1994, vol. 24, no. 4, pp. 234–244.

[63] A. Boukerche, A. Fabbri, and Sajal K. Das, "Analysis of randomized congestion control in DSDV routing," Journal of Parallel and Distributed Computing, vol. 60, no. 8, pp. 65–72, August 2000.

[64] M. Joa-Ng, "A peer-to-peer zone-based two-level link state routing for mobile ad hoc networks," *IEEE J. Sel. Areas Commun.*, vol. 17, no. 8, pp. 1415–1425, Aug. 1999.

[65] S. R. Gopal, R. Chandra, N. S. N. S. V Raos, and S. G. Batsells, "DST—A routing protocol for ad hoc networks using distributed spanning trees," in *Proceedings of IEEE Wireless Communications and Networking Conference (WCNC'99)*, IEEE, New Orleans, Louisiana, 1999, pp. 1543–1547.

[66] X. H. X. Hong, K. X. K. Xu, and M. Gerla, "Scalable routing protocols for mobile ad hoc networks," *IEEE Netw.*, vol. 16, no. 4, pp. 11–21, Aug. 2002.

[67] P. Samar, M. R. Pearlman, and Z. J. Haas, "Independent zone routing: An adaptive hybrid routing framework for ad hoc wireless networks," *IEEE/ACM Trans. Netw.*, vol. 12, no. 4, pp. 595–608, Aug. 2004.

[68] M. Marina, and S. Das, "On-demand multipath distance vector routing in ad hoc networks," in *Proceedings of IEEE Ninth International Conference on Network Protocols*, Riverside, CA, pp. 14–23, 2001.

[69] B. Y. P. Goverde, K. I. M. Taeymans, and K. Lauwers, "Performance study of the better approach to mobile ad hoc networking (B.A.T.M.A.N.) protocol in the context of asymmetric links," in *Proceedings of IEEE International Symposium on World of Wireless Mobile and Multimedia Networks (WoWMoM'12)*, Washington, DC, 2012, pp. 1–24.

[70] R. Torres, L. Mengual, O. Marban, S. Eibe, E. Menasalvas, and B. Maza, "A management Ad Hoc networks model for rescue and emergency scenarios," *Expert Syst. Appl.*, vol. 39, no. 10, pp. 9554–9563, Aug. 2012.

[71] G. Koltsidas, S. Karapantazis, G. Theodoridis, and F. N. Pavlidou, "A detailed study of dynamic manet on-demand multipath routing for Mobile Ad hoc Networks," in *Proceedings of International Conference on Wireless and Optical Communications Networks*, Singapore, 2007, pp. 274–278.

[72] L. E. Quispe, and L. M. Galan, "Behavior of Ad Hoc routing protocols, analyzed for emergency and rescue scenarios, on a real urban area," *Expert Syst. Appl.*, vol. 41, no. 5, pp. 2565–2573, Apr. 2014.

[73] T. T. Son, H. L. Minh, G. Sexton, and N. Aslam, "A novel encounter-based metric for mobile ad-hoc networks routing," *Ad Hoc Netw.*, vol. 14, pp. 2–14, Mar. 2014.

[74] A. K. Gupta, H. Sadawarti, and A. K. Verma, "Review of various routing protocols for MANETs," *Int. J. Inf. Electron. Eng.*, vol. 1, no. 3, pp. 251–259, Jan. 2011.

[75] S. S. Dhillon, X. Arbona, and P. Van Mieghem, "Ant routing in mobile ad hoc networks," in *Proceedings of the Third International Conference on Networking and Services (ICNS'07)*, Athens, Jun. 19–25, 2007.

[76] N.-C. Wang and C.-Y. Lee, "A multi-path QoS multicast routing protocol with slot assignment for mobile ad hoc networks," *Inf. Sci. (Ny).*, vol. 208, pp. 1–13, Nov. 2012.

[77] A. Boukerche, B. Turgut, N. Aydin, M. Z. Ahmad, L. Bölöni, and D. Turgut, "Routing protocols in ad hoc networks: A survey," *Comput. Netw.*, vol. 55, no. 13, pp. 3032–3080, Sept. 2011.

[78] K. Suto, H. Nishiyama, and N. Kato, "Postdisaster user location maneuvering method for improving the QoE guaranteed service time in energy harvesting small cell networks," *IEEE Trans. Veh. Technol.*, vol. 66, no. 10, pp. 9410–9420, Oct. 2017.

[79] S. Cabrero, X. G. Paneda, D. Melendi, R. Garcia, and T. Plagemann, "Using firefighter mobility traces to understand ad-hoc networks in wildfires," *IEEE Access*, vol. 6, pp. 1331–1341, Nov. 2017.

[80] F. Tang, Z. M. Fadlullah, N. Kato, F. Ono, and R. Miura, "AC-POCA: Anticoordination game based partially overlapping channels assignment in combined UAV and D2D-based networks," *IEEE Trans. Veh. Technol.*, vol. 67, no. 2, pp. 1672–1683, Feb. 2018.

[81] D. G. Reina, M. Askalani, S. L. Toral, F. Barrero, E. Asimakopoulou, and N. Bessis, "A survey on multihop ad hoc networks for disaster response scenarios," International Journal of Distributed Sensor Networks, vol. 2015, Article ID 642389, May 2015.

[82] S. V. Mallapur, "Survey on simulation tools for mobile ad-hoc networks," *Int. J. Comput. Netw. Wirel. Commun.*, vol. 2, no. 2, pp. 241–248, Apr. 2012.

[83] N. I. Sarkar, S. Member, and S. A. Halim, "A review of simulation of telecommunication networks: Simulators, classification, comparison, methodologies, and recommendations," Cyber Journals: Multidisciplinary Journals in Science and Technology, Journal of Selected Areas in Telecommunications, vol. 1, no. 3, pp. 10–20, March 2011.

[84] R. Khan, S. M. Bilal, and M. Othman, "A performance comparison of open source network simulators for wireless networks," in Proceedings of the IEEE International Conference on Control System, Computing and Engineering (ICCSCE), pp. 34–38, IEEE, July 2012.

[85] D. G. Reina, S. L. Toral, F. Barrero, N. Bessis, and E. Asimakopoulou, "Modelling and assessing ad hoc networks in disaster *scenarios*," *J. Ambient Intell. Humaniz. Comput.*, vol. 4, no. 5, pp. 571–579, Oct. 2013.

[86] P. C. Ng, and S. C. Liew, "Throughput analysis of IEEE802. 11 multi-hop ad hoc networks," *IEEE/ACM Trans. Netw.*, vol. 15, no. 2, pp. 309–322, May 2007.

[87] S. Ullah, and I. A. Khan, "Survey on MANET routing protocols and challenges". *Wirel. Pers. Commun.*, vol. 112, no. 2, 2257–2294, Aug. 2020.

[88] Kaur, A, "A detailed survey on routing protocols in MANETs". *Wirel. Netw.*, Vol. 26, no. 2, pp. 1295–1320, Sept. 2020.

[89] S. Kumar, and K. Dutta, "Role of ad hoc network in disaster management," *Int. J. Comput. Appl.*, vol. 180, no. 20, pp. 6–9, 2018.

[90] D. G. Reina, S. L. Toral, P. Johnson, and F. Barrero, "A survey on probabilistic broadcast schemes for wireless Ad hoc networks", *Ad Hoc Netw.*, vol. 25, pp. 263–292, Feb. 2015.

[91] W. Dargie, and C. Poellabauer, *Fundamentals of wireless sensor networks: Theory and Practice*, John Wiley & Sons, New York, 2010, p. 336.

Index

Note: **Bold** page numbers refer to tables and *italic* page numbers refer to figures.